HUMAN EVOLUTION

HUMAN EVOLUTION

•

Bernard Wood

A BRIEF
INSIGHT

STERLING

New York / London

www.sterlingpublishing.com

STERLING and the distinctive Sterling logo are registered trademarks of
Sterling Publishing Co., Inc.

Library of Congress Cataloging-in-Publication Data

Wood, Bernard A.
  Human evolution: a brief insight / Bernard Wood.
      p. cm.
  ISBN 978-1-4027-7898-8
  1. Human evolution. I. Title.
  GN281.W67 2011
  599.93'8--dc22
                                                    2010017328

10 9 8 7 6 5 4 3 2 1

Published by Sterling Publishing Co., Inc.
387 Park Avenue South, New York, NY 10016

Published by arrangement with Oxford University Press, Inc.

© 2005 by Bernard Wood
Illustrated edition published in 2011 by Sterling Publishing Co., Inc.
Additional text © 2011 Sterling Publishing Co., Inc.

Distributed in Canada by Sterling Publishing
c/o Canadian Manda Group, 165 Dufferin Street
Toronto, Ontario, Canada M6K 3H6

Book design: Faceout Studio

Please see picture credits on page 166 for image copyright information.

Printed in China
All rights reserved

Sterling ISBN 978-1-4027-7898-8

For information about custom editions, special sales, premium and corporate purchases, please contact
Sterling Special Sales Department at 800-805-5489 or specialsales@sterlingpublishing.com.

Frontispiece: Human evolution exhibit at the American Museum of Natural History, in New York.

CONTENTS

•

ACKNOWLEDGMENTS

•

FOR AN AUTHOR USED TO the luxury of lengthy academic papers and the occasional five-hundred-page monograph, and to the protection afforded by technical language and multiple qualifications, boiling down human evolutionary history to the size constraints and style of a Brief Insight was a considerable challenge. Thanks go to Mark Weiss for advice about genetics, to Matthew Goodrum, Ed Lee, and David Pilbeam for advice about the history of science and of human origins research, to my colleague, Robin Bernstein, to my OUP editor, Marsha Filion, and to an anonymous reviewer, for reading the entire manuscript and making valuable suggestions for revisions. Graduate students in the hominid paleobiology program at George Washington University, and Phillip Williams wittingly and unwittingly contributed by providing information and helping me find "lost" files and notes. I am grateful to several publishers for allowing me to adapt and use previously published images and figures. This book is for my family, young and old, and my teachers.

PEDIGREE OF MAN.

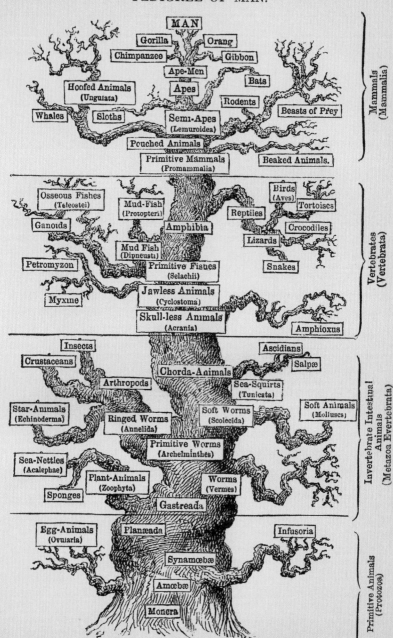

ONE

Introduction

●

MANY OF THE IMPORTANT ADVANCES made by biologists in the past 150 years can be reduced to a single metaphor. All living, or extant, organisms, that is, animals, plants, fungi, bacteria, viruses, are on the surface of an arborvitae, or Tree of Life, and all the types of organisms that lived in the past are situated somewhere on the branches and twigs within that same tree. The extinct organisms on the branches that connect us directly to the root of the tree are our ancestors. The rest, on branches that connect with our own, are closely related to modern humans, but they are not our ancestors.

The "long" version of human evolution would be a journey that starts approximately 3 billion years ago at the base of the TOL with the simplest form of life. We would then pass up the base of the trunk and

This early version of the tree of life is from German biologist Ernst Haeckel's *The Evolution of Man* (1879).

into the relatively small part of the tree that contains all animals, and on into the branch that contains all the animals with backbones. Around 400 million years ago we would enter the branch that contains vertebrates that have four limbs, then around 250 million years ago into the branch that contains the mammals, and then into a thin branch that contains one of the subgroups of mammals called the primates. At the base of this primate branch we are still at least 50–60 million years away from the present day.

The next part of this "long" version of the human evolutionary journey takes us successively into the monkey and ape, the ape, and then the great ape branches of the Tree of Life. Sometime between 15 and 12 million years ago we move into the small branch that gave rise to contemporary modern humans and to the living African apes. Between 11 and 9 million years ago the branch for the gorillas split off to leave just a single slender branch consisting of the ancestors and extinct close relatives of chimpanzees and bonobos (chimps/bonobos) and modern humans. Around 8 to 5 million years ago this very small branch split into two twigs. At the root of the twigs is the most recent common ancestor (MRCA) of chimps/bonobos and modern humans. One of the two twigs ends on the surface of the TOL with the living chimps/bonobos; the other also ends on the surface of the TOL, but with modern humans.

This book focuses on the last stage of the human evolutionary journey, the part between ourselves and the MRCA shared by chimpanzees/bonobos *and* modern humans. Paleoanthropology is the science that tries to reconstruct the evolutionary history of our small twig. It tries to find out how complex, or "bushy," was the small, exclusively human, part of the TOL. But instead of referring to "twigs" we will use the proper

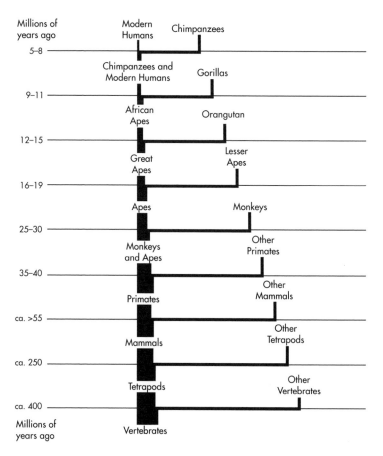

A diagram of the vertebrate part of the Tree of Life
emphasizing the branches that led to modern humans.

biological term *clade*: extinct side branches are called *subclades*. Species
anywhere on the clade that connects the MRCA with modern humans
on the surface of the TOL are called *hominins*; the equivalent species

on the clade that connects the MRCA with chimps/bonobos are called *panins*. And instead of writing out "millions of years" and "millions of years ago" (and the equivalents for thousands of years), we will use instead the abbreviations "myr" and "mya" and "kyr" and "kya."

This introduction has three objectives. The first is to try to explain how paleoanthropologists go about the task of improving our understanding of human evolutionary history. The second is to convey a sense of what we think we know about human evolutionary history, and the third is to try to give a sense of where the major gaps in our knowledge are.

We use two main strategies to improve our understanding of human evolutionary history. The first is to obtain more data. You can get more data by finding more fossils, or by extracting more information from the existing fossil evidence. You can find more fossils from existing sites, or you can look for new sites. You can extract more information from the existing fossil record by using techniques such as laser scanning to make more precise observations about their external morphology. You can also gather information about the internal morphology and biochemistry of fossils. This ranges from using noninvasive imaging techniques such as computed tomography to obtain information about structures like the inner ear, to using a relatively new type of microscopy, confocal microscopy, to investigate the microscopic anatomy of teeth, and using the latest molecular biology technology to determine the structure of small amounts of DNA left in fossils.

The second strategy for reducing our ignorance about human evolutionary history is to improve the ways we analyze the data we do have. These improvements range from more effective statistical methods to the use of novel methods of functional analysis. Researchers also try to improve the ways they generate and test hypotheses about the numbers

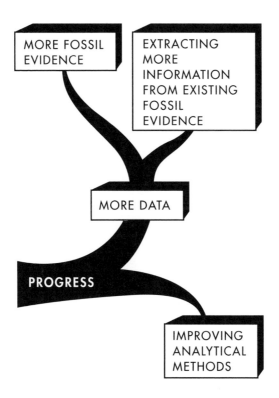

Diagram showing how progress can be made in paleoanthropology research.

of species in the hominin fossil record, and about how those species are related to each other and to modern humans and chimpanzees.

I begin Chapter 2 by reviewing the history of how philosophers and then scientists came to realize that modern humans are part of the natural world. I then explain why scientists think chimpanzees are more closely related to modern humans than they are to gorillas, and why they think the chimp/human common ancestor lived between 8 and 5 mya.

In Chapter 3 I review the lines of evidence that can be used to investigate what the 8–5 myr-old hominin clade looks like. Is it "bushy," or straight like the stem of a thin, spindly plant? How much of it can be reconstructed by looking at variation in modern humans, and what needs to be investigated by searching for, finding, and then interpreting fossil and archaeological evidence? Where do researchers look for new fossil sites, and how do they date the fossils they find? In Chapter 4 I explain how researchers decide how many species there are within the hominin clade. I also review the methods researchers use to determine how many hominin subclades there are, and how they are related to one another.

In Chapter 5 I consider "possible" early hominins. The chapter reviews four collections of fossils that represent each of the "candidate" taxa that have been put forward for being at the very base of the hominin clade. Then in Chapter 6 I look at "archaic" and "transitional" hominins. These are fossil taxa that almost certainly belong to the hominin clade, but which are still a long way from being like modern humans; most of these are likely to be close relatives, but not our ancestors. Chapter 7 looks at hominins researchers believe might be the earliest members of the genus *Homo*: we call these "premodern" *Homo*. I look at the earliest fossil evidence of premodern *Homo* from Africa, and then follow *Homo* as it moves out of Africa into the rest of the Old World.

Chapter 8 considers evidence about the origin and subsequent migrations of anatomically modern humans, or *Homo sapiens*. When and where do we find the earliest fossil evidence of anatomically modern humans? Did the change from premodern *Homo* to anatomically modern humans happen several times and in several different regions of the world? Or did anatomically modern humans emerge just once, in one place, and then spread out, either by migration or by interbreeding, so that modern humans eventually replaced regional populations of premodern *Homo*?

Finally, what will not be in this book? This Brief Insight to "Human Evolution" will concentrate on the physical and not the cultural aspects of human evolution. The latter, often referred to as "Prehistoric Archaeology," is the topic of a book called *Prehistory: A Very Short Introduction*.

TWO

Finding Our Place

●

LONG BEFORE RESEARCHERS BEGAN to accumulate material evidence about the many ways modern humans resemble other animals, and long before Charles Darwin and Gregor Mendel laid the foundations of our understanding of the principles and mechanisms that underlie the connectedness of the living world, Greek scholars had reasoned that modern humanity was part of, and not apart from, the natural world. When did the process of using reason to try to understand human origins begin, and how did it develop? When was the scientific method first applied to the study of human evolution?

The Sicilian writer Empedocles (ca. 490–430 BCE) suggested that during one phase of the great cycle of transformations that make up the

Classical Greek and Roman thinkers considered the use of tools, fire, and verbal language crucial components of humanity. Pictured here are pictographs of handprints from a cave in Patagonia, South America.

history of the universe, random mixings of the fundamental elements of nature (for him, there were only four such elements—earth, air, fire, and water) occasionally produced separate, isolated parts of animals, an arm here, a leg there, and that these occasionally met and bonded together to form yet more complicated beings, and sometimes this happened in ways that were fitted to survive and form creatures like modern humans and the other animals. Weird as this idea was, it did intriguingly combine the ideas of random, nonteleological mixture, and fitness for survival under the pressures of natural selection. But Empedocles' ideas attracted little support, and both Plato and Aristotle rejected him as fundamentally mistaken. The Roman philosopher Lucretius, writing in the first century BCE, proposed that the earliest humans were unlike contemporary Romans. He suggested that human ancestors were animal-like cave dwellers, with neither tools nor language. Both classical Greek and Roman thinkers viewed tool and fire making and the use of verbal language as crucial components of humanity. Thus, the notion that modern humans had evolved from an earlier, primitive form appeared early on in Western thought.

Reason Is Replaced by Faith

After the collapse of the Roman Empire in the fifth century, Greco-Roman ideas about the creation of the world and of humanity were replaced with the narrative set out in Genesis: reason-based explanations were replaced by faith-based ones.

The main parts of the narrative are well-known. God created humans in the form of a man, Adam, and then a woman, Eve. Because they were the result of God's handiwork Adam and Eve must have come equipped with language and with rational and cultured minds. According to this

This early illustration of Adam and Eve (top left) is one of the scenes in this folio depicting the history of Adam in the Ashburnham Pentateuch, a sixth-century Spanish illuminated manuscript of the first five books of the Old Testament.

version of human origins, the first humans were able to live together in harmony, and they possessed all the mental and moral capacities that, according to the biblical narrative, set humanity above and apart from other animals.

The biblical explanation for the different races of modern humans is that they originated when Noah's offspring migrated to different parts of the world after the last big biblical flood, or deluge. The Latin for "flood" is diluvium, so we call anything very old "antediluvial," or dating from "before the flood." Explanations for the creation of the living world involving successive floods had implications for the science that was to become known as paleontology. All the animals created after a flood must inevitably perish at the time of the next flood. Thus "antediluvial" animals should never coexist with the animals that replaced them. We will return to this and other implications of diluvialism later in this chapter.

The Bible also has an explanation for the rich variety of human languages. It suggests that God wanted to promote confusion among the people constructing the Tower of Babel, and that she or he did so by creating mutually incomprehensible languages. In the Genesis version of human origins, the Devil's successful temptation of Adam and Eve in the Garden of Eden forced them and their descendants to learn afresh about agriculture and animal husbandry. They had to reinvent all the tools needed for civilized life.

With very few exceptions, Western philosophers living in and immediately after the Dark Ages (fifth to twelfth centuries) supported a biblical explanation for human origins. This changed with the rediscovery and rapid growth of natural philosophy that was only later called science. But, paradoxically, not long after the scientific method began to be applied to the study of human origins in the nineteenth and twentieth centuries, some religious groups responded to attempts by scientists to interpret the Bible less literally by being even stricter about their biblical literalism. This reaction was the origin of creationism, and of what, erroneously, is called "creation science."

During the Dark Ages very few Greek classical texts survived in Europe. The few that did survive were read and valued by Muslim philosophers and scholars, and some of them were translated into Arabic. When the Muslims were finally driven out of Spain in the fifteenth century, a few medieval Christian scholars were curious enough to translate these manuscripts from Arabic into Latin. Some of these translated texts dealt with the natural world, including human origins. For example, the thirteenth-century Italian Christian philosopher, Thomas Aquinas, integrated Greek ideas about nature and modern humans with some of the Christian interpretations based on the Bible. The work of Thomas Aquinas and his contemporaries laid the foundations of the Renaissance, the period when science and rational learning were reintroduced into Europe.

Saint Thomas Aquinas is shown here in a panel from a large altarpiece in the church of San Domenico in Ascoli, Italy.

Science Reemerges

The move away from reliance on biblical dogma was especially important for those who were interested in what we now call the natural sciences,

The first-edition title page of Bacon's *Instauratio magna* (1620), of which the *Novum organum* was a part.

such as biology and the earth sciences. An Englishman, Francis Bacon, was a major influence on the way scientific investigations developed. Theologians use the deductive method: beginning with a belief, they then deduce the consequences of that belief. Bacon suggested that scientists should work in a different way he called the "inductive" method. Induction begins with observations, also called evidence or "data." Scientists devise an explanation, called a "hypothesis," to explain those observations. Then they test the hypothesis by making more observations, or in sciences like chemistry, physics, and biology, by conducting experiments. This inductive way of doing things is the way the sciences involved in human evolution research are meant to work.

Bacon summarized his suggestions about how the world should be investigated in aphorisms, and set these out in his book called the *Novum Organum or True Suggestions for the Interpretation of Nature*, published in 1620. His message was a simple one. Do not be content with reading about an explanation in a book. Go out, make observations, investigate the phenomenon for yourself, then devise and test your own hypotheses.

Anatomy Starts to Become Scientific

Nearly three-quarters of a century before Bacon published this advice, a major change had already occurred in anatomy, the natural science closest to the study of human evolution. That change was the work of Andreas Vesalius. Born in 1514 in what is now Belgium, Vesalius finished his medical studies in 1537, the year he was appointed to teach anatomy and surgery in Padua, Italy.

This illustration of Vesalius performing a dissection in a crowded anatomical theater appeared on the title page of his *De humani corporis fabrica libri septem* (1543).

Vesalius's own anatomy education was typical for the time. The professor sat in his chair (hence professorships are called "chairs"), at a safe distance from the human body that was being dissected by his assistant, and he (they were all men) read out loud from the only locally available textbook. It did not take long for Vesalius to realize that he and his fellow students were being told one thing by their professor, and were being shown something different by the professor's assistant. In 1540 Vesalius visited Bologna, where, for the first time, he was able to compare the skeletons of a monkey and a human. He realized that the textbooks used by his professors were based on a confusing mixture of human, monkey, and dog anatomy, so he resolved to write his own, accurate, human anatomy book. The result, the seven-volume *De Humani Corporis Fabrica Libri Septem*, or "On the Fabric of the Human Body," was published in 1543. Vesalius performed the dissections and sketched the drafts of the illustrations: the *Fabrica* is one of the great achievements in the history of biology. Vesalius's successful efforts to make anatomy more rigorous ensured that scientists would henceforth have access to reliable information about the structure of the human body.

Geology Emerges

Another field of science relevant to the eventual study of human origins, geology (now usually referred to as "earth science"), developed more gradually than anatomical science. One of the implications of interpreting the Genesis narrative literally is that the world, and therefore humanity, cannot have had a long history. There is a long tradition of biblically based chronologies, beginning with people like Isidore of Seville and the Venerable Bede in the sixth and seventh centuries, respectively. The one cited most often was published in 1650 by James Ussher, then archbishop of

Armagh in Ireland. He used the number of "begats" in the Book of Genesis to calculate the precise year of the act of Creation, which, according to his arithmetic, was in 4004 BCE. Subsequently another theologian, John Lightfoot, of Cambridge University, England, refined Ussher's estimate and declared that the act of Creation took place precisely at 9 A.M. on October 23, 4004 BCE.

The development of geology was substantially influenced by the Industrial Revolution. The excavations involved in making "cuttings" for canals and railroads gave amateur geologists the opportunity to see previously hidden rock formations. Pioneer geologists such as Smith and Hutton paved the way for Charles Lyell to set out a rational version of the history of the earth in *The Principles of Geology* published in 1830. Lyell's book influenced many scientists, including Charles Darwin, and it helped establish fluvialism and uniformitarianism as alternatives to biblically based diluvial explanations for the state of the landscape. Fluvialism suggested that erosion by rivers and streams had reduced the height of mountains and created valleys and thus played a major role in shaping the contours of the earth. Uniformitarianism suggested the processes that had shaped the earth's surface in the past, such as erosion and volcanism, were the same processes we see in action today. Lyell also championed the principle that rocks and strata generally increase in age the farther down they are in any relatively simple geological sequence. Barring major and obvious upheavals and deliberate burial, the same principle must apply to any fossils or stone tools contained within those rocks. The lower in a sequence of rocks a fossil or a stone tool is, the older it is likely to be.

The implications of the new science of geology were profound. There was no need to invoke the biblical floods or divine intervention to explain the appearance of the earth, and the work of practitioners like William Smith and James Hutton provided an alternative calendar, which suggested that the earth and its inhabitants must be substantially older than the six thousand years suggested by Ussher's Genesis-based scenario.

Fossils

Classical Greek and Roman writers had recognized the existence of fossils but they mostly interpreted them as remnants of the ancient monsters that figure prominently in their myths and legends. By the eighteenth century geologists began to accept that lifelike structures in rocks were the remains of extinct animals and plants, and that there was no need to invoke supernatural reasons for their existence. The association of the fossil evidence of exotic extinct animals with creatures closely related to living forms in the same strata effectively refuted the diluvial theory, for the latter does not allow for any mixing of modern and ancient, or antediluvial, animals.

In addition to the important conclusions reached by pioneer geologists about the long history of the earth, several other factors influenced seventeenth- and eighteenth-century scientists to consider alternatives to the Genesis account of human origins. Explorers were returning from distant lands with eyewitness accounts of modern humans living in crude shelters, using simple tools, and existing by hunting and gathering. This was so far from the state of humanity in their homeland that European travelers described the people they observed as living in a state of "savagery." According to the Genesis narrative, no human beings created by God should be living in such a state.

A Catalogue of Life

The same explorers and traders who had returned to Europe with tales of the behavior of primitive people also brought back descriptions and sometimes suitably preserved specimens of exotic plants and animals. When these discoveries were added to the more familiar plants and animals from Europe, they made for a perplexing array of plant and animal life. The living world badly needed a system for describing and organizing it. Several schemes were put forward, notably one by John Ray, who introduced the concept of the species. However, the one that has stood the test of time was devised by a Swede called Karl Linné, a name we know better in its Latinized form, Carolus Linnaeus.

Carolus Linnaeus (1707–78), who created the seven-level system for naming organisms, the basis for the binomial nomeclature used today, is shown here in Lapland dress, holding a plant, with instruments hanging at his waist.

Classification schemes try to group similar things together in increasingly broad, or inclusive, categories. Think of the following example of a classification of automobiles. It has seven levels, or categories; it begins with the most inclusive category and ends with a small, exclusive, group. The levels are "Vehicles," "Powered Vehicles," "Automobile," "Luxury Car," "Rolls-Royce," "Silver Shadow," and "1970 Silver Shadow II." The Linnaean classification system also recognizes seven basic levels. The most inclusive category, the equivalent of "Vehicles" in our example, is the kingdom, followed by the phylum, class, order, family, genus, with the species being the smallest, least inclusive, formal category. Linnaeus's original seven-level system has been expanded by adding the category "tribe" between the genus and family, and by introducing the prefix super- above a category, and the prefixes sub- and infra-, below it. These additions increase the potential number of categories below the level of order to a total of twelve.

The groups recognized at each level in the Linnaean hierarchy are called "taxonomic groups." Each distinctive group is called a "taxon" (pl. "taxa"). Thus, the species *Homo sapiens* is a taxon, and so is the order Primates. When the system is applied to a group of related organisms, the scheme is called a Linnaean taxonomy, usually abbreviated to a "taxonomy." The Linnaean taxonomic system is also known as the binomial system because two categories, the genus and species, make up the unique Latinized name (e.g., *Homo sapiens* = modern humans; *Pan troglodytes* = chimpanzees) we give to each species.

You can abbreviate the name of the genus, but not the species. So you can write *H. sapiens* and *P. troglodytes*, but not *Homo s.* or *Pan t.*, as there can sometimes be more than one species name in that genus that begins with the same first letter, such as *Homo sapiens* and *Homo soloensis*.

Evidence of Connections

Trees are common metaphors. In religion, for example, in Christianity, the Great Chain of Being is sometimes represented as a tree. Modern humans are on top of the tree, with other living animals placed within the tree at heights corresponding to their level of complexity. However, in contemporary life sciences the Tree of Life is not a metaphor: it is taken more literally. In a modern scientific Tree of Life the relative size of the part of the tree given over to any particular group of living things reflects the number of taxa in that group, and the pattern of branching within the tree reflects the way scientists think plants and animals are related.

When the first science-based Trees of Life were constructed in the nineteenth century, the closeness of the relationship between any two animals had to be assessed using morphological evidence that could be studied with the naked eye or with a conventional light microscope. The assumption was that the larger the number of shared structures, the closer their branches will be within the TOL. Developments in biochemistry during the first half of the twentieth century meant that, in addition to this traditional morphological evidence, scientists could use evidence about the physical characteristics of molecules. The earliest attempts to use biochemical information for determining relationships used protein molecules found on the surface of red blood cells and in plasma. Both these lines of evidence emphasized the closeness of the relationship between modern humans and chimpanzees.

Proteins are the basis of the machinery that makes other molecules, like sugars and fats, and ultimately the tissues that make up the components of our bodies, such as muscles, nerves, bones, and teeth. In 1953 James Watson and Francis Crick, with the help of Rosalind Franklin, discovered that the nature of proteins, the building blocks of our bodies, is determined by the details of a molecule called DNA (short for deoxyribose nucleic acid). Scientists have shown since that DNA transmitted from parents to their offspring contains coded instructions, called the genetic code. This, in large measure, determines what the bodies of those offspring will look like. These developments in molecular biology meant that instead of working out how species are related by comparing traditional morphology, or by looking at the morphology of protein molecules, scientists could in theory determine relationships by comparing the DNA that dictates the structure and shape of proteins, and ultimately what we look like.

When these methods, first traditional anatomy, then the morphology of protein molecules, and finally the structure of DNA (the details of how DNA is compared are given below) were applied to more and more of the organisms in the TOL, it became apparent that animal species that were similar in their anatomy also had similar molecules and similar genetic instructions. Researchers have also shown that, even though the wing of an insect and the arm of a primate look very different, the same basic instructions are used during their development. This is additional compelling evidence that all living things are connected within a single TOL. The only explanation for these similarities and this connectedness that has withstood scientific scrutiny is evolution, and the only mechanism for evolution that has withstood scientific scrutiny is natural selection.

Evolution—An Explanation for the Tree of Life

Evolution means gradual change. In the case of animals this usually (but not always) means a change from a less complex animal to a more complex animal. We now know that most of these changes occur during speciation, which is when an "old" species changes quite rapidly into a "new," different, species. Although the Greeks were comfortable with the idea that the behavior of an animal could change, they did not accept that the structure of animals, including modern humans, had been modified since they were spontaneously generated. Indeed Plato championed the idea that living things were unchanging, or immutable, and his opinions influenced philosophers and scientists until the middle of the nineteenth century.

A French scientist, Jean Baptiste Lamarck, in his *Philosophie Zoologique* published in 1809, set out the first scientific explanation for the TOL. In the English-speaking world Lamarck's ideas were popularized in an influential book called *Vestiges of the Natural History of Creation* (1844). We know that *Vestiges* influenced the two men, Charles Darwin and Alfred Russel Wallace, who, independently, hit upon the concept that the main mechanism driving evolution was natural selection.

Charles Darwin's contributions to science did not include the idea of evolution; that notion antedated Charles Darwin. What Darwin did contribute was a coherent theory about the way evolution could work. As we will see, Darwin's theory of natural selection accounts for both the diversity and the branching pattern of the TOL. Other books that influenced Darwin's thinking were Robert Malthus's *Essay on the Principle of Population* (1798) and Charles Lyell's *Principles of Geology*. Malthus stressed that resources are finite and this suggested to Darwin that imbalances between the resources available and the demand for them might be the driving force behind the selection needed to make evolution happen. Lyell's fluvial explanation for the evolution of the surface of the earth was much like the gradual morphological change that Darwin suggested was responsible for the modification of existing species to produce new ones. Darwin was also goaded into action by the work and philosophy of William Paley. Paley was a champion of the notion that animals were so well adapted for their habitat that this cannot have been due to chance. He suggested that they must have been designed, and if so there must be a designer, and that the designer must have been God. Paley's creationist interpretations provoked Darwin to think about an alternative.

Charles Darwin is pictured here in a photograph from ca. 1854, about five years before he published *The Origin of Species.*

Charles Darwin made three seminal contributions to evolutionary science. The first was the recognition that no two individual animals are alike: they are not perfect copies. Darwin's related contribution was the idea of natural selection. In a nutshell, natural selection suggests that, because resources are finite, and because of random variation, some individuals will be better than others at accessing those resources. That variant will then gain enough of an advantage that it will produce more surviving offspring than other individuals belonging to the same species. Biologists refer to this advantage as an increase in an animal's "fitness." Darwin's notebooks are full of evidence about the effectiveness of the type of artificial selection used by animal and plant breeders. Darwin's genius was to think of a way that the same process could occur naturally. Darwin's third contribution was to make the notion of a Tree of Life a reality, and not just a metaphor.

Selection, and thus evolution, will work only if, in the case of natural selection, the offspring of a mating faithfully inherits the feature, or features, that confer(s) greater genetic fitness. What Darwin did not realize (nor for that matter did any other prominent contemporary biologist) was that while he was putting the finishing touches to the *Origin of Species*, the genetic basis of variation and the essential rules of inheritance were being painstakingly worked out in a monastery garden in Brno, in what is now the Czech Republic.

The Flowering of Genetics

The discipline of genetics was established on the basis of deductions made by Gregor (this was his Augustinian monastic name; his original forename was Johann) Mendel about the collection of artificially bred pea plants he maintained in the garden of his monastery. Mendel presented the results

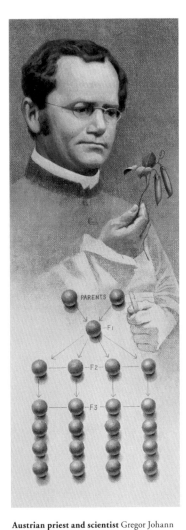

Austrian priest and scientist Gregor Johann Mendel is shown with a diagram summarizing the results of his experiments with pea plants.

of his breeding experiments to the Natural Science Society in Brno in 1865, but he did not use the terms *gene* (meaning the smallest unit of heredity) or *genetics*. The word *gene* was not coined until 1909, nine years after Mendel's pioneering experiments came to the notice of evolutionary scientists. It was Mendel's good fortune that his various plant-breeding experiments provided several examples of a simple one-to-one link between a gene and a trait—these are called single-gene, or "monogenic," effects.

Mendel's simple dichotomies, yellow or green, smooth or wrinkled, are called "discontinuous" variables. Paleoanthropologists have to deal with "continuous" variables such as the size of a tooth, or the thickness of a limb bone. These have smooth, curved, distributions, not the neat columns that result from data like Mendel's. How do you get continuous curves from discontinuous columns of data? The answer is that many genes are involved in determining the size of a tooth, or the thickness of a limb bone, so that what looks like a curve is in reality the combination of many columns.

Our Closest Relatives

Not so long ago a book on human origins would have devoted a substantial number of pages to descriptions of the fossil evidence for primate evolution. This was in part because it was assumed that at each stage of primate evolution one of the fossil primates would have been recognizable as the direct ancestor of modern humans. However, we now know that for various reasons many of what we once thought were ancestors are highly unlikely to be ancestral to living higher primates. Instead, this account will concentrate on what we know of the evolution and relationships of the great apes. It will review how long Western scientists have known about the great apes, and it will show how ideas about their relationships to each other, and to modern humans, have changed. It will also explore which of the living apes is most closely related to modern humans.

Among the tales of exotic animals brought home by explorers and traders were descriptions of what we now know as the great apes, that is, chimpanzees and gorillas from Africa, and orangutans from Asia. Aristotle referred to "apes" as well as to "monkeys" and "baboons" in his *Historia animalium* (literally the "History of Animals"), but Aristotle's "apes" were the same as the "apes" dissected by the early anatomists. They were almost certainly short-tailed macaque monkeys from North Africa.

One of the first people to undertake a systematic review of the differences between modern humans and the chimpanzee and gorilla was Thomas Henry Huxley. In an essay entitled "On the Relations of Man to the Lower Animals" that formed the central section of his 1863 book called *Evidence as to Man's Place in Nature*, he concluded the anatomical differences between modern humans and the chimpanzee and gorilla were less marked than the differences between the two African apes and the orangutan. Darwin used this evidence in his *The Descent of Man*, published in 1871, to suggest that, because the African apes were morphologically closer to modern humans than to the only great ape known from Asia, the ancestors of modern humans were more likely to be found in Africa than elsewhere. This deduction played a critical role in pointing most researchers toward Africa as a likely place to find human ancestors. As we will see in the next chapter, those who considered the orangutan our closest relative looked to Southeast Asia as the most likely place to find modern human ancestors.

Developments in biochemistry and immunology during the first half of the twentieth century allowed the search for evidence about the nature of the relationships between modern humans and the apes to be shifted from traditional morphology to the morphology of molecules. The earliest attempts to use proteins to determine primate relationships were made just after the beginning of the twentieth century, but it was not until the early 1960s that the first results of a new generation of analyses were reported. The famous US biochemist Linus Pauling coined the name "molecular anthropology" for this area of research. Two reports, both published in 1963, provided crucial evidence. Emile Zuckerkandl, another pioneer molecular anthropologist, used enzymes to break up the protein hemoglobin from blood red cells into its peptide components, and when he separated them using a small electric current, the patterns made by the peptides from a modern human, a chimpanzee, and a gorilla were indistinguishable. The second contribution was by Morris Goodman, who has spent his life working on molecular anthropology. He used techniques borrowed from immunology to study samples of a serum (serum is what is left after blood has clotted) protein called albumin taken from modern humans, apes, and monkeys. He concluded that the albumins of modern humans and chimpanzees were so alike he could not tell them apart.

Proteins are made up of a string of amino acids. In many instances one amino acid may be substituted for another without changing the function of the protein. In the 1960s and 1970s Vince Sarich and Allan Wilson, two Berkeley biochemists interested in primate and human evolution, exploited these minor variations in protein structure in order to determine their evolutionary history, and therefore, presumably, the evolutionary history of the taxa being sampled. They, too, concluded that modern humans and the African apes were very closely related.

Interrogating the Genome

The discovery of the chemical structure and significance of the DNA molecule meant that affinities between organisms could be pursued by comparing their DNA, or genome. This potentially eliminated the need to rely on morphology, be it traditional anatomy or the morphology of proteins, for information about relatedness. Now, instead of using proxies researchers could study relatedness directly by comparing DNA. The DNA within the cell is located either within the nucleus as nuclear DNA, or within organelles called mitochondria in mtDNA. In DNA sequencing the base sequences of each animal are determined and then compared.

Sequencing methods have been applied to living hominoids and the number of studies increases each year. The genomes of several modern humans and a few chimpanzees have been sequenced. Information from

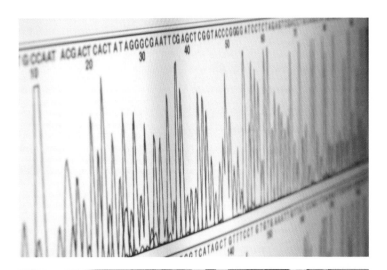

Maps like this chromatograph (top) showing a DNA sequence plot the position of certain known genes and are preliminary to sequencing the entire genome. Even before these DNA maps can be created, clinicians must process and label the DNA samples themselves, as shown in this recent photograph (above) from the National Institutes of Health.

both nuclear and mtDNA suggests that modern humans and chimpanzees are more closely related to each other than either is to the gorilla. When these differences are calibrated using the "best" paleontological evidence for the split between the apes and the Old World monkeys, and if we assume that the DNA differences are neutral so that more DNA difference means more time has elapsed since the clades split, the prediction is that the hypothetical ancestor of modern humans and chimps/bonobos lived between 8 and 5 mya. When other, older, calibrations are used, the predicted date for the split is somewhat older (e.g., >10 mya).

Implications for Interpreting the Human Fossil Record

The results of recent morphological analyses of both skeletal and dental anatomy, and the anatomy of the soft tissues such as muscles and nerves, are also consistent with the compelling DNA evidence that chimps/bonobos are closer to modern humans than they are to gorillas. But some attempts to use the type of traditional morphological evidence that is conventionally used to investigate relationships among fossil hominin taxa did not find a particularly close relationship between modern humans and chimpanzees. Instead, chimpanzees clustered with gorillas.

This has important implications for researchers who investigate the relationships among hominin taxa. Either they need to use types of information (e.g., more precise measurement of their shape) about skulls, jaws, and teeth that *are* capable of confirming the close relationship between chimps and modern humans, or they need to find other sources of morphological evidence, such as information about the shape of the limb bones, and see if those data are capable of recovering the relationships among living higher primates supported by the DNA evidence.

Table 1. A Traditional Taxonomy (A) and a Modern Taxonomy (B) That Take Account of the Molecular and Genetic Evidence that Chimpanzees Are More Closely Related to Modern Humans Than They Are to Gorillas: Extinct Taxa Are in Bold Type

A. Superfamily Hominoidea (hominoids)

Family Hylobatidae (hylobatids)
Genus *Hylobates*

Family Pongidae (pongids)
Genus *Pongo*
Genus *Gorilla*
Genus *Pan*

Family Hominidae (hominids)
Subfamily Australopithecinae (australopithecines)
   Genus *Ardipithecus*
   Genus *Australopithecus*
   Genus *Kenyanthropus*
   Genus *Orrorin*
   Genus *Paranthropus*
   Genus *Sahelanthropus*
Subfamily Homininae (hominines)
   Genus *Homo*

B. Superfamily Hominoidea (hominoids)

Family Hylobatidae (hylobatids)
Genus *Hylobates*

Family Hominidae (hominids)
Subfamily Ponginae (pongines)
    Genus *Pongo*
Subfamily Gorillinae (gorillines)
    Genus *Gorilla*
Subfamily Homininae (hominines)
    Tribe Panini (panins)
        Genus *Pan*
    Tribe Hominini (hominins)
    Subtribe Australopithecina (australopiths)
        Genus *Ardipithecus*
        Genus *Australopithecus*
        Genus *Kenyanthropus*
        Genus *Orrorin*
        Genus *Paranthropus*
        Genus *Sahelanthropus*
    Subtribe Hominina (hominans)
        Genus *Homo*

THREE

Fossil Hominins:
Their Discovery and Context

•

AS EXPLAINED IN CHAPTER 1, a hominin is the label we give to anatomically modern humans and all the extinct species on, or connected to, the modern human twig of the Tree of Life. In this chapter I discuss what the hominin fossil record consists of, how it is discovered, and how it and its context are investigated.

The Hominin Fossil Record

A fossil is a relic or trace of a former living organism. Only a tiny fraction of living organisms survive as fossils, and until people were buried deliberately, this also applied to hominins. We are almost certain that the fossils that do survive are a biased sample of the original population, and

Primate skulls, apes and humans.

I discuss the implications of this in more detail in the next chapter. Fossils are usually, but not always, preserved in rocks. Scientists recognize two major categories of fossils. The smaller category, trace fossils, includes footprints, like the 3.6 myr-old footprints from Laetoli in Tanzania that I discuss in Chapter 6, and coprolites (fossilized feces). The larger category, true fossils, consists of the actual remains of animals or plants. In the hominin fossil record they so outnumber trace fossils that when we use the word *fossil* it will normally apply to true fossils. Animal fossils usually consist of the hard tissues such as bones and teeth. This is because hard tissues are more resistant to being degraded than are soft tissues such as skin, muscle, or the gut. Soft tissues are only preserved in the later stages of the hominin fossil record: for example, the Bog People found in Denmark and elsewhere in Europe.

Fossilization

The chances that an individual early hominin's skeleton would have been preserved as part of the fossil record are very small. Carnivores, such as the predecessors of modern lions, leopards, and cheetahs, would most likely have had the first pick at the carcass of a dead hominin. After them would have come the terrestrial scavengers, led by hyenas, wild dogs, and smaller cats, then birds of prey, then insects, and finally bacteria. Within two to three years—a surprisingly short time—between them these organisms are capable of removing most traces of any large mammal.

For the bones and teeth of a dead hominin to be preserved as fossils, they would need to have been covered quickly by silt from a stream, by sand on a beach, or by soil washed into a cave. This protects the prospective fossil from further degradation and allows fossilization to take place. Fossilization of a bone begins when chemicals from the surrounding sediments replace the organic material in the hard tissues; later on, chemicals begin to replace the inorganic material in bones and teeth. These replacement processes proceed for many years, and in this way a bone turns into a fossil. Fossils are essentially bone- or tooth-shaped rocks. In the meantime the sediments that surround the fossil are themselves being converted into rock.

Diagenesis is the word scientists use to describe all the changes that occur to bones and teeth during fossilization. Fossils from different sites, and even fossils from different parts of the same site, show different degrees of fossilization because of small-scale differences in their chemical environment. When fossils are preserved in hard rocks, and when they are freshly exposed, the fossils are very durable. However, if a fossil is exposed to erosion by wind and rain for any length of time, or if it has been softened by the acids that are produced by plant roots, fossil bone can be as fragile as wet tissue paper. In these cases researchers have to infiltrate the fragile bone with liquid plastic, or its equivalent, in order to stop the fossil from disintegrating further. Obviously, deliberate burial greatly increases the chance that skeletons will be preserved in good condition. It is one of the main reasons why the human fossil record gets so much better about 60–70 kya.

Most hominin fossils are found in rocks formed from sediments laid down by rivers, or on lakeshores, or in the floors of caves. Generally older rocks (and thus the fossils they contain) are in the lower layers, and the younger ones are nearer the surface: this is called the law of superposition. However, relative movement of rocks brought about by tension and compression, such as the shearing that occurs along faults in the earth's crust, can confound this general principle. Sedimentary rocks that form in caves are also prone to being jumbled up in even more complex ways.

Water that percolates down from the surface can soften and then dissolve old sediments. This produces Swiss-cheese-like cavities, which are then filled by more recent sediments. So within caves new sediments may be below old ones.

Earth scientists use the appearance, texture, and distinctive chemistry of rocks to describe and classify them. For example, they might refer to one layer as a "pink tuff," or another as a "silty-sand." Just as there are rules for naming new species, there are rules and conventions for naming the strata of a newly discovered sedimentary sequence, and there is the equivalent of a Linnaean taxonomy for rocks.

The layer of rock a fossil was buried in is referred to as its "parent horizon." Hominin fossils found within a particular rock layer are, unless there is obvious evidence that they were deliberately buried, considered to be the same age as that layer. A fossil found embedded in a rock is described as being found "in situ." Most hominin fossils, however, have been displaced through erosion from their parent horizon; these are called "surface finds." In order to reliably connect a surface find to its parent horizon, it helps if the fossil still has some of the parent rock, or matrix, attached to, or embedded in, it. This is why careful scientists never completely clean the matrix from a fossil.

Finding Fossil Hominins

Where do paleoanthropologists look for early hominin fossils? In the nineteenth century Charles Darwin argued that, because the closest living relatives of modern humans, chimpanzees and gorillas, were both confined to Africa, then it was probable that the common ancestor of modern humans was also likely to have lived in Africa. So, for the past seventy-five years, and especially the last fifty years, Africa has been a focus of human origins field research. But researchers cannot possibly search all of Africa. Are there particular places where hominin fossils are likely to be found?

Paleoanthropologists look where rocks of the right age (say back to 10 mya) have been exposed by natural erosion. Erosion occurs in places where streams and rivers cut through rock layers. This is especially likely in regions where the earth's crust has been buckled and cracked as large landmasses, called tectonic plates, are pushed together. The floor and walls of rift valleys are formed when the area between major cracks, or faults, is forced downward, and the earth's crust on the outside of the major faults is thrust upward. The faults associated with rift valleys are sometimes so deep that the liquid core of the earth escapes through them. When it is under very high pressure, the molten core escapes as in a volcanic eruption; otherwise, it "leaks" slowly as a flow of molten lava. Some volcanic eruptions consist of ash (called tephra), which is rich in the chemicals potassium and argon. Rocks formed from these ash layers are called tuffs, and tuffs are the way many East African hominin fossil sites are dated. Tuffs also have a distinctive chemical profile, or "fingerprint," and this allows geologists to trace a single tuff not only within a large fossil site, but also across many hundreds of kilometers from one site to another. Sometimes hot volcanic ash falls not on the land but on water, and the holes in the lumps of the volcanic pumice people buy for the bathroom are caused by the air bubbles that form when hot ash falls on water.

Fossils are exposed on the sides and floors of the sort of valleys that form as streams and rivers erode their way through the blocks of sediment that are thrown up at faults. Locations like these are called "exposures," and the places on these exposures where fossils have been found are called "localities." In East Africa scientists look for hominin fossils in rocks of the right age that have been exposed by the combination of volcanic activity, called tectonism, and erosion in and around the rift valley. Olduvai Gorge, in Tanzania, is probably the best-known example of a rift valley site where tectonism has caused rocks to slide past each other, and where subsequent erosion has exposed rocks that contain hominin fossils.

Early hominin fossils are found in a very different geological context in southern Africa. Here, they are found in caves that form when rain runs through cracks in the limestone. Small cracks expand into big cracks, big cracks become cavities, and cavities coalesce to become caves that fill with soil washed in from the surface. Leopards use the trees that grow in the entrances of the caves as a place to hide carcasses, and hyenas use the entrances of such caves as a den. Scientists think that most of the hominin fossils found in the southern African caves were taken there by leopards or hyenas, or by bone-collecting animals such as porcupines.

C. K. (Bob) Brain demonstrating the complex stratigraphy at Swartkrans, one of the southern African cave sites where early hominin remains have been found.

Although Africa is the major focus of fieldwork today, it was not that way until well into the twentieth century. Before that time the search for human fossils was conducted in Europe and Asia. Europe was where the first prehistorians lived and worked, so it is to be expected that they would have taken advantage of any opportunity that presented itself in their own region before looking for the fossil remains of our ancestors in more exotic places. Just as in 1871 Charles Darwin predicted that Africa would be the birthplace of humankind, Ernst Haeckel, a prominent German naturalist, in 1874 suggested that the presence of the orangutan, the only non-African great ape, in what was then called the Dutch East Indies (now Borneo, Java, and Sumatra in Indonesia) made that region a likely birthplace for humanity. Two years before the publication of Haeckel's influential book, the naturalist Alfred Russel Wallace (1872) had included detailed information about the morphology and the habits of the orangutan in his book about the natural history of the Malay Archipelago.

Haeckel's logic and perhaps Wallace's vivid descriptions of the orangutan evidently appealed to a young trainee surgeon, Eugène Dubois, for in the late 1880s he took a job in the region so he could look for human ancestors. His most famous find, the top of a brain case of a creature that had brow ridges unlike any seen on modern humans, was recovered in 1891 in the bank of the Trinil River in Java. Not all the human ancestors discovered in Asia were found in sediments cut into by rivers. The famous Peking Man fossils came from a cave at a site now called Zhoukoudian, near Beijing in China.

Teamwork

The teams that nowadays look for hominin fossils in Chad, Ethiopia, Libya, or Eritrea must include a wide range of experts. In addition to paleoanthropologists, geologists, dating experts, and paleontologists who can identify and interpret the fossil remains of the animals and plants found with the hominins, a multidisciplinary team should include experts on the factors that bias the fossil record, and may also include earth scientists who can interpret the chemistry of the soils in order to reconstruct ancient habitats. The team's members have to travel to remote and sometimes dangerous places where they, along with local hired workers who help search for and excavate fossils, need supplies of water, food, and fuel. Leaders of expeditions must have good organizational skills in addition to their scientific qualifications. Big expeditions to inaccessible Central and East African fossil sites are expensive to mount, with the largest ones having annual budgets of many thousands of dollars. The southern African cave sites are mostly much more accessible. The majority lie within an hour's journey time by car from Johannesburg or from Pretoria. This enables scientists to supervise research while working in universities and museums in nearby cities.

Fossils Rediscovered

Some dramatic hominin fossil discoveries are made in museums. It is always worthwhile going through the collections of "nonhuman" fossils recovered from a hominin fossil site. Even the best paleontologists can miss things as they sort through hundreds of bone fragments. In the past when important hominin discoveries were made they were sometimes sent away to experts for their assessment, and unless great care is taken, specimens can be muddled or mislabeled. For example, records show that when a remarkably complete skeleton of a Neanderthal baby was recovered from the site of Le Moustier it was sent to Marcellin Boule for an assessment of its age. However, all trace of the skeleton seemed to have been lost until a researcher found the bones of a neonate among the stone tools from the site of Les Eyzies! Fortunately, some of the bones were still in their original matrix, and this matched rocks in the Vezere River, which runs past Le Moustier.

Dating Hominin Fossils

Geologists can usually work out the temporal sequence of fossils within a small fossil site. But how do you compare the ages of fossils found at localities hundreds of kilometers apart, and how do you compare the ages of fossils from sites on different continents? To answer these questions you need ways of dating the fossils. Traditionally dating methods have been divided into two categories, absolute and relative.

Absolute dating methods are mostly applied to the rocks in which the hominin fossil was found, or to nonhominin fossils recovered from the same horizon. Researchers must take great care to preserve the evidence that links a fossil to a particular rock layer. Absolute dating methods rely on knowing the time it takes for natural processes, such as atomic decay, to run their course, or they relate the fossil horizon to precisely calibrated global events such as reversals in the direction of the earth's magnetic field. This is why absolute dates can be given precisely in calendar years. The best known of these absolute dating methods, radiocarbon dating, is appropriate only for the later stages of human evolution. After 5,730 years (plus or minus 40 years) half of the carbon 14 (^{14}C) there was when the organism died has been converted to nitrogen 14 (this is why this length of time is called its "half-life"). Radiocarbon dating has been used successfully for dating *H. sapiens* fossils from Australia and Europe, but radiocarbon dates older than 40 ky are unreliable because the amounts of radiocarbon left are too small to be measured precisely.

Most of the hominin fossils from East African sites such as Olduvai Gorge in Tanzania, Koobi Fora in Kenya, and Hadar in Ethiopia are from horizons sandwiched between layers of volcanic ash, or tephra, that are rich in isotopes of potassium and argon. Because radioactive potassium and argon convert (or decay) into their daughter products more slowly than carbon 14, potassium/argon and argon/argon dating methods can be used on rocks that contain fossils and stone tools from the early (older than ca. 300 ky) part of the hominin fossil record.

Paleomagnetic dating uses the complex record of reversals of the direction of the earth's magnetic field caused by currents in the liquid

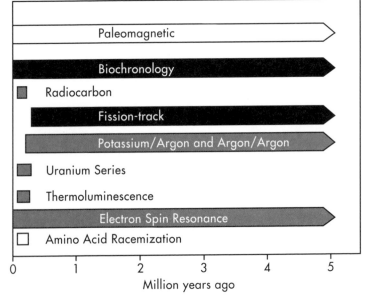

Some of the methods used to date fossil hominins and the time periods they cover. (Stanford, Craig; Allen, John S.; Anton, Susan C., *Biological Anthropology*, 2nd, ©2009. Electronically reproduced by permission of Pearson Education, Inc., Upper Saddle River, New Jersey.)

core of the earth. For long periods in its history the direction of the earth's magnetic field has been the exact opposite of what it is now. The contemporary direction is called "normal" and the opposite one "reversed." When the suspended particles settle prior to the formation of a sedimentary rock, minute amounts of magnetic metal in the particles mean that each of them behaves like a magnet. When they settle they line up with the direction of the earth's magnetic field at the time, and give the rock as a whole a detectable magnetic direction, or "polarity." Researchers compare the sequence of changes in magnetic direction preserved in the hominin fossil-bearing sediments with the magnetic record preserved in cores taken from the floor of the deep ocean (called paleomagnetic columns) and try to find the best match. Some sequences are seen more than once in the reference column, so it helps if another absolute dating method can be used to show researchers which part of the paleomagnetic record they should focus on. A long period of paleomagnetic stability is called a "chron," and a relatively short-lived change in magnetic field direction within a chron is called a "subchron." Olduvai Gorge was the first early hominin site to be dated using magnetostratigraphy, and when subchrons were named and not numbered as they are now, one of them was called the "Olduvai Event."

Another group of absolute dating methods called amino acid racemization dating uses biochemical reactions as a clock. For example, eggshell contains an amino acid called leucine. When a shell is formed initially all the leucine is in the L-form. However, over time this L-form of leucine converts, or racemizes, at a more or less steady rate to an alternate version, called the D-form. Thus, the ratio of the two forms, plus the rate of conversion, provides a date for when the shell was formed. Many later African hominin fossil sites contain fragments of ostrich eggshell, and if we make the reasonable assumption that the eggshell in a horizon is the same geological age as any hominins it contains, then ostrich eggshell (OES) dating can provide a potentially useful method. Ostrich eggshell dating is one of several methods (others are electron spin resonance, ESR, and uranium series dating, USD) scientists use to date hominin fossil sites that are between the ranges of radiocarbon and potassium argon dating. These methods are particularly useful for dating sites between 300 and 40 kya.

Relative dating methods mostly rely on matching nonhominin fossils found at a site with equivalent evidence from another site that has been reliably dated using absolute methods. If the animal fossils found at Site A are similar to those at Site B, Site A can be assumed to be approximately the same age as Site B. Compared to absolute dating methods, relative dating methods provide only approximate ages for fossils. The use of animal remains for dating, called "biochronology," has been especially important for dating early hominin fossils from the southern African cave sites. Nearly all of these sites contain antelope and monkey fossils. Because the same animals have been absolutely dated at key East African sites, researchers can apply these dates to the layers that contain equivalent fossils in the southern African caves. Biochronology has also been used to date hominin fossil sites in Chad and at Dmanisi, in Georgia.

Dendrochronology, the use of tree rings for relative dating, has been used to improve the precision of carbon dating. Annual tree rings are so reliable that they have been used to correct carbon dates that have been affected by recent human-induced, or anthropogenic, changes in levels of carbon isotopes in the atmosphere.

Reconstructing Past Environments

Just as the contours of the earth's surface are different than they were several million years ago, past environments in a region are not necessarily the same as those we see today. Researchers reconstruct past environments using geological and paleontological evidence. Chemical analysis is used to tell whether a soil was laid down in moist or dry conditions. Paleontologists can tell a lot about the paleohabitat from the types of animal fossils that are found along with the fossil hominins. They use both large mammals and small micromammals (such as mice and gerbils) to reconstruct past environments. Small micromammals are especially useful because their geographical ranges are more restricted than those of larger mammals, so they are likely to provide more precise habitat reconstructions. Fossilized owl pellets (these are made of the food owls regurgitate to feed their young) are a good source of information about micromammals because owls hunt small mammals within a relatively small range. Researchers who use larger mammals like primates to reconstruct past environments have to be careful not to assume that the habitat preferences of the ancestors were like those of their modern-day representatives. For example, although modern colobus monkeys are mainly leaf eaters who live in dense woodland, their ancestors lived in more open habitats, so the presence of colobus monkeys at a 5 myr-old site does not mean the same as finding contemporary colobus monkeys.

Global Climate Change

Hominin evolution has taken place at a time when there have been major changes in world climate. Researchers study climate change by looking at deep-sea cores. Microscopic organisms called foraminifera (usually shortened to "forams") are suspended in the water of the world's oceans. These foraminifera take up two forms of oxygen isotope: one of them, oxygen 16 (^{16}O), is lighter; the other, oxygen 18 (^{18}O), is heavier. When global temperatures are higher more of the lighter oxygen evaporates, so the ratio of the light to the heavy form reduces: the opposite occurs when global temperatures are cooler. Researchers use the proportions of the two oxygen isotopes to track the temperature of the oceans, and they use ocean water temperature as a proxy for global climate. But the climate in a region is the result of a complex interaction between global climate and local influences such as latitude, altitude, and the presence of mountain ranges.

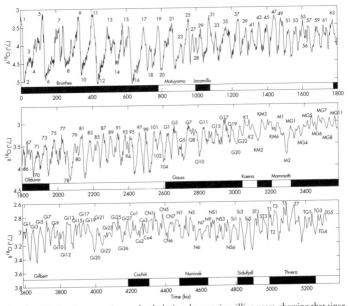

Plot of oscillations in oxygen isotope levels during the past six million years, showing that since 3 mya the global climate has shown a general cooling trend. (Lorraine E. Lisiecki and Maureen E. Raymo, "A Pliocene-Pleistocene Stack of 57 Globally Distributed Benthic δ^{18}O Records," *Paleoceanography* 20, PA1003 (2005). Copyright 2005 American Geophysical Union. Reproduced/modified by permission of American Geophysical Union.)

During the period from 8 to 5 mya the earth experienced the beginning of a long-term drying and cooling trend. Early hominin evolution took place in Africa at the time of these climatic changes, and the possible influence of climate change on the origin of the hominin lineage will be explored further in Chapter 5. Later in hominin evolution cyclical changes in global climate, measured using deep-sea cores, were superimposed on the long-term cooling trend. Prior to 3 mya global climate was subject to 23 kyr hotter/drier and cooler/wetter cycles. Around 3 mya the periodicity of these cycles switched to 41 kyr, and around 800 kya it switched yet again to a 100 kyr cycle. These 100 kyr cycles are the ones responsible for the periods of intense cold recorded in the northern hemisphere during the past million years. These long cycles had another important impact on human evolution because when so much ice is locked up in the icecaps at the north and south poles, it is inevitable that the sea level will fall. This would have exposed much of what we call the continental shelf. Reductions in sea level of this magnitude allowed modern human ancestors to migrate from the Old World to both Australasia and the New World.

I. QUADRUPEDIA.

Corpus hirfutum. *Pedes* quatuor. *Femina* viviparæ, lactiferæ.

II. AVES.

Corpus plumofum. *Alæ* duæ. *Pedes* bini. *Femina* oviparæ.

QUADRUPEDIA

ANTHROPO-MORPHA.	Homo.	Nofce te ipfum.	H { Europæus albefc. / Americanus rubefc. / Afiaticus fufcus. / Africanus nigr.
	Simia.	ANTERIORES. POSTERIORES. / Digiti 5. 5. / Poteriores anterioribus fimiles.	Simia cauda carens. Papio. Satyrus. Cercopithecus. Cynocephalus.
	Bradypus.	Digiti 3. vel 4 . . . . 3.	Ai. Ignavus. Tardigradus.
FERÆ. *Dentes primores 6, utrinque: intermedii longiores: omnes acuti. Pedes multifidi, unguiculati.*	Urfus.	Digiti 5. . . . . . 5. Scandens. / Mamma 4. (Abd.) / Calcaneis infiftit. Pollen extus pofitus.	Urfus. Coati Merg. Wickhenzi Angl.
	Leo.	Digiti 5. . . . . . 4. Scandens. / Mamma 2. ventrales. / Lingua aculeata.	Leo.
	Tigris,	Digiti 5. . . . . . 4. Scandens. / Mamma 4 umbilicales. / Lingua aculeata.	Tigris. Panthera.
	Felis,	Digiti 5. fc. 4. pect. 4. abdom. Scandens. / Mamma 8. fc. 4. pect. 4. abdom. / Lingua aculeata.	Felis. Catus. Lynx.
	Muftela,	Digiti 5. . . . . . 5. Scandens. / Dentes molares 4. utrinque.	Martes. Zibellina. Vivera. Muftela. Putorius.
	Didelphis.	Di_vi . . . . / Mamma 8. intra burfulam abdomin.	Philander, Pofum.
	Lutra.	Digiti 5. . . . . . 5. Palmipes.	Lutra.
	Odobænus.	Digiti 5. . . . . . 5. Palmipes. / Dentes intermedii fuperiores longifi.	Rofi. Morfus.
	Phoca.	Digiti 5. . . . . . 5. Palmipes. / Mamma duæ umbilicales.	Canis marinus.
	Hyæna.	Cullum fuperne jubatum. / Cauda brevia.	Hyæna Veter. Vivam Londini nuper vidi & defcripfit ARYAD.
	Canis,	Digiti 5. . . . . . 4. / Mamma 10. fc. 4. pect. 6. abdom.	Canis. Lupur. Squillachi. Vulpes.
	Meles.	Ungues medii digiti ipfi longiores. / Corpus fuperne albicat: inferne nigricat.	Taxus. Zibetha.
	Talpa.	Digiti 5. . . . . . 5. anteriores maximi.	Talpa.
	Erinaceus,	Digiti 5. . . . . . 5. / Spinis vel lorica fquamofa munitus.	Echinus terreftris. Armadillo.
	Vefpertilio.	Digiti 5. . . . . . 5. / Pes anticus in alam expanfus. / Mamma 2. pectorales.	Vefpertilio. Felis volans Seb. Canis volans Seb. Glis volans Seb.
GLIRES. *Dentes primores 2. utrinque.*	Hyftrix.	Aures humanæ. Corpus fpinofum.	Hyftrix.
	Sciurus.	Digiti 4. . . . . . 5. / Cauda longiffima laniger.	Sciurus. . . . volans.
	Caftor,	Digiti 5. . . . . . 5. palmipes poftice. / Cauda horizontalis, plana, nuda.	Fiber.
	Mus.	Digiti 4. . . . . . 5. / Cauda teres, fquamofa, hirfuta.	Rattus. Mus domefticus. . . . bahericus. . . . marmorus. Lemures. Marmota.
	Lepus.	Digiti 5. . . . . . 5. / Cauda breviffima, villofa.	Lepus. Cuniculus.
	Sorex.	Digiti 5. . . . . . 5. / Dentes canini adfunt.	Sorex.
JUMENTA. *Dentes primores incifi, obtu_*	Equus.	Mamma 2. inguinales. / Pedes integri.	Equus. Afinus. Onager. Zebra.
	Hippopotamus.	Mamma 2. inguinales. (Arifi.) / Pedes quadrifidi.	Equus marinus.
	Elephas.	Mamma 2. pectorales. / Pedes 5. callis inftructi.	Elephas. ? Rhinoceros.
	Sus.	Mamma 10. abdominales. / Pedes biungulati : raro fimplices.	Sus. Aper. Porcus. Barbyrouffa.

AVES

ACCIPITRES. *Rofrum uncinatum.*	Pfittacus.	Digiti pedis anticl 2. poftici 2.
	Strix.	Digiti pedis anticl 3. poftici 1 quorum extimus retrorfum.
	Falco.	Digiti pedis anticl 3. poftici 1.
PICÆ. *Rofrum fuperne compreffum, convexum.*	Paradifæa.	Penna 2. longiffimæ, fingularæ nec alia, nec uropygio infident.
	Coracias.	Pes 4dact. Rectrices exteriores g breviores.
	Corvus.	Pes 4dact. Rectrices æquales.
	Cuculus.	Digiti pedis anticl 2. poftici 2. / Rofrum læve.
	Picus.	Digiti pedis anticl 2. poftici 2. / Rofrum angulatum.
	Certhia.	Pes 4dact. Rofr. gracile incurv
	Sitta.	Pes 4dact. Rofr. triangulare.
	Upupa.	Pes 4dact. Caput plumis crifta
	Ifpida.	Pes 4dact. cujus digitus extimu adnectitur tribus articulis.
MACRORHYN.CHÆ. *Rofr. longit. acut.*	Grus.	Caput criftatum.
	Ciconia.	Ungues plani, fubrotundi.
	Ardea.	Ungues medius inferne ferratus.
ANSERES. *Os dentato-ferratum.*	Platelea.	Rofr. depreffo-planum, apice fu
	Pelecanus.	Rofr. depreffum, apice unguicu inferne burfa inftructum.
	Cygnus.	Rofr. conico-coavexum.
	Anas.	Rofr. conico-depreffum.
	Mergus.	Rofr. cylindriforme, apice acu
	Graculus.	Rofr. conicum, apice aduncu.
	Colymbus.	Rofr. fubulatum. Pedes infra æq
	Larus.	Rofr. fubulatum. Pedes in æquili
SCOLOPACES. *Rofrum cylindraceo-teretiufculum.*	Hæmatopus.	Pes 3dact. Rofri apex compref
	Charadrius.	Pes 3dact. Rofri apex teres.
	Vanellus.	Pes 4dact. Rofrum digitis brevi Caput pennato criftatum.
	Tringa.	Pes 4dact. Rofrum digitis brevi Caput fimplex.
	Numenius.	Pes 4dact. Rofrum digitis longi
	Fulica.	Pes 4dact. Digiti membranis aut Caput carnofo-criftatum.
GALLINÆ. *Rofrum conico-incurvum.*	Struthio.	Pes 2dact. abfque poftico.
	Cafuarius.	Pes 3dact. abfque poftico. Caput galea & palearibus cru
	Otis.	Pes 3dact. abfque poftico. Caput fimplex.
	Pavo.	Pes 4dact. Caput corolla pennac.
	Meleagris.	Pes 4dact. Frons papilla. Gula brand unici longitudine ftructus.
	Gallina.	Pes 4dact. Frons membrana carn Gula membr. duplici longitudi

FOUR

Fossil Hominins:
Analysis and Interpretation

•

PALEOANTHROPOLOGISTS USE MANY METHODS to work out the significance of newly discovered hominin fossil evidence. The hominin fossils must be assigned to a taxon, or taxa, the taxa must be classified, their relationships to other fossil and living taxa worked out, and their behavior reconstructed.

Classification and Taxonomy

Western science classifies all living things according to a scheme devised in 1758 by the Swedish naturalist Carolus Linnaeus. The basic unit of the scheme is the species, a group of morphologically similar animals

Table of the Animal Kingdom (Regnum Animale) from Carolus Linnaeus's first edition (1735) of *Systema naturae*.

that consistently breed productively with each other. Individual living animals all belong to a species, similar species are grouped into genera, genera are grouped into tribes, tribes into families, and so on up to categories like kingdoms. Modern humans, *Homo sapiens*, belong in the species *H. sapiens*, the genus *Homo*, and the tribe Hominini.

A subdiscipline of classification, called "nomenclature," is devoted to prescribing how names should be used in the Linnaean system. There is a formal code for regulating nomenclature, and scientists who think they have discovered a new species must follow this code. Rules in the code govern the types of name that can be given to a new species or genus. For example, the names of commercial products are prohibited: *Burgerking ipodensis* would not be an acceptable binomial for a new hominin species. It is also important to make sure that the name of an existing taxon is not inadvertently used for a new taxon; otherwise they will be confused.

When researchers decide to introduce a new species, they have to choose one fossil as its "type" specimen. Usually a relatively well-preserved fossil is selected from among those found at the time of the initial discovery: it does not have to be a typical (i.e., an average) member of the species. The significance of the type specimen is that the taxon name is irrevocably attached to it. So, for example, if the type specimen of *Homo neanderthalensis* was found to be different from all the other fossils included in *H. neanderthalensis*, then they would have to be assigned to a new species, and it would need to be given a new name. The name *H. neanderthalensis* cannot be used independently of the type specimen; where it goes, the name goes, too. If researchers eventually decide that a particular specimen should be moved to a new species, then it takes its species name with it. Age counts in nomenclature: if two type specimens end up in the same species, the oldest name is the one that has to be used.

A species is an example of a taxon. All the Linnaean categories are taxa, but when researchers write about "a taxon" they are usually referring to a species. How species are arranged in an increasingly inclusive hierarchy (i.e., larger and larger clusters of species) is called a taxonomy, literally a "scheme for taxa." Taxonomic analysis is the process of determining what taxon hominin fossils should be put in. First, researchers have to decide whether a newly found fossil belongs in an existing hominin taxon. Only if they are convinced that it cannot be assigned to an existing species can they begin to think of making a new species with a new name. The same principles apply all the way up the Linnaean hierarchy, so researchers should establish a new genus only if they are convinced the new species cannot be accommodated in any of the existing hominin genera, and so on up the Linnaean hierarchy.

Taxonomic analysis and the other methods of analysis described below are based on a detailed assessment of the morphology of a fossil. Its morphology, or phenotype, is what the fossil looks like, both externally and internally. Morphology can be gross morphology, which is what the eye can see unaided, or microscopic morphology, which is what can be seen with a variety of types of microscope. Researchers prepare detailed qualitative descriptions of the size and shape of the fossil, but they also try to capture that information in the form of measurements as a quantitative description. In its simplest form quantitative descriptions consist of distances between defined anatomical landmarks on the fossil: these are called linear measurements. Laser beams and other technologies borrowed from medical imaging now allow researchers to capture details of the external morphology and the internal structure of fossils much more precisely than in the past. For example, Glenn Conroy, a paleoanthropologist, and Charles Vannier, a medical imaging specialist, both from

Washington University in St. Louis, pioneered the use of computerized tomography (or CT) imaging to study the internal structure of a fossil hominin cranium from Taung in southern Africa. Subsequently Frans Zonneveld, a medical imaging specialist from Utrecht, and Fred Spoor, a paleoanthropologist from University College London, further developed these methods so that they can now provide information about the cavities that contain the part of the inner ear that registers movement and sounds. Researchers use these data to help sort hominin fossils into species and to reconstruct their posture and hearing.

Researchers must be sure the measurements made on fossils accurately reflect the size and shape of the bone or tooth before it was fossilized. Bones and teeth crack if they are exposed to daily cycles of heating and cooling. Rock matrix gets inside the cracks and artificially enlarges the dimensions of a bone or tooth. Likewise, if a fossil bone is exposed on the surface of the ground in dry and windy conditions both before and after fossilization, sand grains carried by the wind have a "sandblasting" effect and remove part of the outer layer of cortical bone. This erosion artificially reduces the size of the fossil bone. The measurements and the nonmetrical morphology of a newly recovered fossil are compared with those of similar specimens in existing fossil taxa. Closely related living animals (in the case of hominins this means modern humans and the African apes) are usually used as models to help decide how much variation should be tolerated within a single species. But Cliff Jolly, a primatologist from New York University who has spent thirty years studying what happens at the boundary between distinctive groups of baboons, suggests that baboons and their close relatives are in some ways a better analogue for hominin evolution. He points out that not only are baboons more widespread than chimpanzees and gorillas, but they are also similar to hominins with respect to the pattern and timing of their recent evolutionary history.

Reconstructing Whole Fossils from Fragments

Hominin fossils several millions of years old are seldom found in good condition. The brain case and the face are particularly fragile and are easily trampled by hoofed animals and crushed by rocks falling from the roof of a cave. Sometimes just one fragment of the brain case is all that is left of a cranium. In a few cases more is preserved, but if the pieces are tiny it is a challenge to reassemble them. It is like a three-dimensional jigsaw puzzle with lots of sky and no clouds, and with no picture on the front of the box to help you. One option is to painstakingly reassemble the pieces by hand, but this can take hundreds of hours even by a skilled anatomist who knows every detail of a skull.

Marcia Ponce de León and Christoph Zollikofer from the Anthropological Institute of Zurich, and Phillip Gunz from the Max Plank Institute for Evolutionary Anthropology in Leipzig, are experts in a new research area called "virtual anthropology." They have used computer power and advances in software design to devise an alternative to reassembling hominin fossils by hand. The fossil is scanned using a laser, and a "virtual" version is displayed on the computer screen. Researchers can then move and rotate each piece in any direction to see if any of the pieces fit. The software also enables a missing piece on one side of the cranium to be replaced by mirror imaging the equivalent piece from the other side. Zollikofer and Ponce de León have recently used these methods to make a virtual reconstruction of the cranium of *Sahelanthropus tchadensis*, a potential early hominin. Similar software in conjunction with CT scans enables structures buried deep in the bone, like the air sinuses, the bony canals of the inner ear, or the roots of the teeth, to be seen clearly.

Determining Age and Sex

Even if one has a complete or nearly complete skeleton, determining the sex and developmental age of hominin fossil remains can be difficult. These difficulties are compounded when all that remains are small fragments of a cranium. The age at death of a fossil individual that has finished growing is difficult to determine precisely. Dental development can help determine the age of immature individuals, but once all the teeth are erupted and the roots of the teeth are formed dental evidence is less useful.

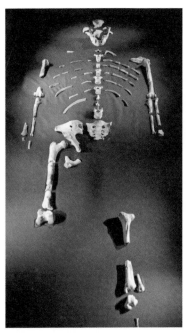

The size and shape of the bones and teeth, the extent of muscle markings, and the size and shape of the pelvis (although pelvic fragments are rare in the hominin fossil record) are the usual ways the sex of an individual fossil is determined. The underlying assumption is that because in many nonhuman primates

The *Australopithecus afarensis* fossil nicknamed Lucy, the bone fragments of which are shown in this photograph from 2007, was identified as female based on the shape of her pelvis, which is similar to that of female *Homo sapiens*, and her small size in relation to other *A. afarensis* fossils.

males are larger than females, then early hominin males were also likely to have been larger than early hominin females. This is one aspect of sexual dimorphism, a term that refers to all the differences among individuals that are related to their sex. However, when you are dealing with a sparse fossil record, overall size is not always a reliable guide to sex.

There are also complications if one unthinkingly extrapolates modern human sexual dimorphisms to early hominins. For example, in modern humans many pelvic sex dimorphisms occur because of compromises between the requirements of bipedalism and the need in modern human females for space in the pelvis to give birth to a large-brained infant. The same dimorphisms, however, might not apply to small-brained early hominins who are not bipedal in the same way that modern humans are: their pelves may show a unique pattern of sexual dimorphism.

Species and Species Identification

The most widely used scientific definition of a species is the biological species concept (BSC) that is linked with the late Ernst Mayr, a distinguished Harvard evolutionary biologist. This suggests that a species is a "group of interbreeding natural populations, reproductively isolated from other such groups." This is all well and good when you can observe living animals, and check who is mating with whom, but it is self-evident that this method will not work when we try to recognize species in the fossil record. However, because members of the same species mate with each other and not with members of another species, they resemble each other more closely than they do individuals belonging to any other species. Thus, in the absence of information about its mating habits, we can use the appearance, structure, and (if any DNA is preserved) the genetic makeup of an individual fossil, to help allocate it to a species.

But there are problems when researchers try to apply these methods to the fossil record. The first difficulty is that we do not have complete animals in the hominin fossil record. It is customary to divide the components of animals into two categories, soft tissues, such as muscles, nerves, arteries, and hard tissues, such as bones and teeth. The fossil record for human ancestors is restricted to the remains of the hard tissues, and many of these are just fragments of bones and teeth. So the problem for paleoanthropologists is how to assign a fossil to a species when the only evidence you have is several worn and broken teeth, or a piece of jaw, or part of a thigh bone. It is like trying to identify the make and model of a car when all you have is a front brake disc, a nondescript piece of the gearbox, and a rear lightbulb!

The second problem is time. Each species has a history, with a beginning (speciation), a middle, and an end. Species either die out without leaving any direct descendants (extinction), or they become the common ancestor of one or more new "daughter" species. The average fossil mammal species lasts for between one and two million years. During such a long history the appearance of that species is unlikely to stay the same. Random variation and morphological responses to climatic variation will cause it to change. But as long as its members mate only with members of the same species, then the species should continue to be distinctive. However, even if a scientist spends his or her whole career observing just one living species the scientist will have studied that species for just an instant during its existence. So the variation you see in museum collections of skeletons belonging to a modern species that have been collected over the course of a hundred years, or so, is not an appropriate model for deciding how much variation one should tolerate in a sample made up of fossils collected at sites that span several hundred thousand years of time.

A good analogy is of a running race. A fossil is like a single still photograph of a long-distance running race. But a long-lived species may well be sampled several times during its history. Paleoanthropologists need to work out ways of telling whether they are looking at several photographs of the same running race, or single photographs of several different running races. In the case of human evolution this means looking at collections of modern human, and higher primate skeletons, and then using the size and shape variation within those living taxa as a guide to how much variation researchers should tolerate within a collection of fossils assigned to a single species. If the variation is less than that seen in the living taxa, then there are good reasons to conclude that only one species is represented in the collection of fossils. Because of the extra time involved with fossil samples, paleoanthropologists try to make an educated guess about the amount of variation they are prepared to tolerate in their fossil sample before they declare that the variation is "too great" to be contained in a single species. But it is still only a guess, albeit a well-educated and scientifically informed guess.

Deciding how many species are represented in a collection of early hominin fossils is made more difficult because biological variation among hominins, including fossil hominins, is continuous. Therefore, where the boundaries between fossil taxa are drawn is a matter of legitimate scientific judgement and debate. The discovery of new fossils or the introduction of new analytical methods often means that boundaries have to change, or paleoanthropologists have to reconsider the utility of their categories and labels. A new species should be established only if there are really good grounds for believing the new fossil evidence does not belong to an existing species. There needs to be even stronger evidence to establish a new genus.

Speciation

Some researchers think that new species are the result of gradual change involving the whole population. This interpretation of speciation is called "phyletic gradualism," and the form of speciation associated with it is known as "anagenesis." Others see speciation as the result of bursts of rapid evolutionary change concentrated in a geographically restricted subset of the population. This interpretation of speciation is called the "punctuated equilibrium" model. In the latter model in the long interval between the periods of rapid evolutionary change there should be no sustained trends in the direction of morphological evolution, just "random walk" fluctuations in morphology. Species formation in that mode is called "cladogenesis" and the term "stasis" is used to describe the periods of morphological stability between speciation episodes. Almost all researchers now accept that most of the morphological change involved in evolution occurs at the time of speciation. In some circumstances speciation may be due to quite large-scale changes in the genotype brought about by rearrangements in the chromosomes. Researchers have suggested that this may have been the mechanism underlying speciation in higher primates.

Particularly intensive periods of species generation and diversification are called "adaptive radiations." They tend to be associated with an opportunity to exploit a new environment, or when extinctions in other groups mean that adaptive opportunities become available in an existing environment. At times like these some lineages tend to generate more species than others, and they are referred to as being "speciose."

All species, including modern humans, will ultimately become extinct. What is at issue is whether extinctions are determined by the

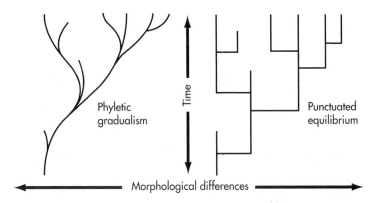

The two main hypotheses "phyletic gradualism" and "punctuated equilibrium" about the timing of the morphological change that occurs during evolution. (Miller, Barbara D.; Wood, Bernard; Balkansky, Andrew; Mercader, Julio; Panger, Melissa, *Anthropology*, 1st, ©2006. Adapted and electronically reproduced by permission of Pearson Education, Inc., Upper Saddle River, New Jersey.)

intrinsic properties of a species, or by extrinsic factors such as changes in the environment, or by a combination of the two. These competing hypotheses can be tested in the laboratory by varying the conditions under which rapidly evolving organisms such as fruit flies are kept. They can also be investigated by comparing the fossil record with independent evidence about changes in past climates.

Splitters and Lumpers

The taxonomy used in this Brief Insight recognizes a relatively large number of hominin species, but not all researchers recognize that many species. Researchers who subscribe to taxonomies that recognize many species are called "splitters." Those who recognize fewer species are called "lumpers." Both groups of researchers are looking at the same evidence; they just interpret it differently. Most disagreements among

Splitters separate *H. rudolfensis* (fossil reconstruction at left) and *H. habilis* (fossil reconstruction at right) into two species, whereas lumpers consider both to be *H. habilis*.

paleoanthropologists about how many species to recognize in the human fossil record are due to differences in how they interpret variation. Researchers who stress the importance of continuities within the fossil record generally opt for fewer species, whereas those who stress discontinuities within the fossil record will generally recognize more species. However, when all is said and done, all taxonomies are hypotheses. If scientists explain their taxonomy, then other scientists can reinterpret the evidence in any way they choose, as long as everyone makes it clear which fossil specimens they are allocating to the species taxa they choose to recognize.

Table 2. Two Taxonomic Hypotheses, One "Splitting" and One "Lumping," for the Hominin Fossil Record

Informal Group	Splitting Taxonomy	Age (myr)	Type Specimen	Main Fossil Sites
Possible hominins	S. tchadensis	7.0–6.0	TM 266-01-060-1	Toros-Menalla, Chad
	O. tugenensis	6.0	BAR 1000'00	Lukeino, Kenya
	Ar. ramidus	5.7–4.3	ARA-VP-6/1	Gona and Middle Awash, Ethiopia
	Ar. kadabba	5.8–5.2	ALA-VP-2/10	Middle Awash, Ethiopia
Archaic hominins	Au. anamensis	4.2–3.9	KNM-KP 29281	Allia Bay and Kanapoi, Kenya
	Au. afarensis	4.0–3.0	LH 4	Belohdelie, Dikika, Fejej, Hadar, Maka, and White Sands, Ethiopia; Allia Bay, Tabarin, and West Turkana, Kenya
	K. platyops	3.5–3.3	KNM-WT 40000	West Turkana, Kenya
	Au. bahrelghazali	3.5–3.0	KT 12/H1	Bahr el ghazal, Chad
	Au. africanus	3.0–2.4	Taung 1	Gladysvale, Makapansgat [Mb 3 and 4], Sterkfontein [Mb 4], and Taung, South Africa
	Au. garhi	2.5	BOU-VP-12/130	Bouri, Ethiopia
	P. aethiopicus	2.5–2.3	Omo 18.18	Omo Shungura Formation, Ethiopia; West Turkana, Kenya; Laetoli; Tanzania
	P. boisei	2.3–1.3	OH 5	Konso and Omo Shungura Formation, Ethiopia; Chesowanja, Koobi Fora, and West Turkana, Kenya; Melema, Malawi; Olduvai and Peninj (Natron), Tanzania

Continued

Table 2. *continued*

Informal Group	Splitting Taxonomy	Age (myr)	Type Specimen	Main Fossil Sites
	P. robustus	2.0–1.5	TM 1517	Cooper's, Drimolen, Gondolin, Kromdraai [Mb 3], and Swartkrans [Mbs 1, 2, and 3], South Africa
Premodern *Homo*	*H. habilis*	2.4–1.6	OH 7	Omo Shungura Formation, Ethiopia; Koobi Fora, Kenya; ?Sterkfontein and ?Swartkrans, South Africa; Olduvai, Tanzania
	H. rudolfensis	2.4–1.6	KNM-ER 1470	Koobi Fora, Kenya; Uraha, Malawi
	H. ergaster	1.9–1.5	KNM-ER 992	?Dmanisi, Georgia; Koobi Fora and West Turkana, Kenya
	H. erectus	1.8–0.2	Trinil 2	Many sites in the Old World e.g., Melka Kunturé, Ethiopia; Zhoukoudian, China; Sambungmacan, Sangiran, and Trinil, Indonesia; Olduvai, Tanzania
	H. floresiensis	0.095–0.018	LB1	Liang Bua, Flores, Indonesia
	H. antecessor	0.7–0.5	ATD6-5	Gran Dolina, Atapuerca
	H. heidelbergensis	0.6–0.1	Mauer 1	Many sites in Africa and Europe, e.g., Mauer, Germany; Boxgrove, England; Kabwe, Zambia
	H. neanderthalensis	0.2–0.03	Neanderthal 1	Many sites in Europe, the Near East, and Asia
Modern *Homo*	*H. sapiens*	0.2–pres	None designated	Many sites in the Old World and some in the New World

Informal Group	Lumping Taxonomy	Age (myr)	Taxa Included from the Splitting Taxonomy
Possible hominins	*Ar. ramidus*	7.0–4.5	*Ar. ramidus, Ar. kadabba, S. tchadensis, O. tugenensis*
Archaic hominins	*Au. afarensis*	4.2–3.0	*Au. afarensis, Au. anamensis, Au. bahrelghazali, K. platyops*
	Au. africanus	3.0–2.4	*Au. africanus*
	P. boisei	2.5–1.3	*P. boisei, P. aethiopicus, Au. garhi*
	P. robustus	2.0–1.5	*P. robustus*
Premodern *Homo*	*H. habilis*	2.4–1.6	*H. habilis, H. rudolfensis*
	H. erectus	1.9–0.018	*H. erectus, H. ergaster, H. floresiensis*
Modern *Homo*	*H. sapiens*	0.7–pres	*H. sapiens, H. antecessor, H. heidelbergensis, H. neanderthalensis*

Cladistic Analysis

Once the taxonomy of a new discovery has been worked out, researchers move on to the next stage. This involves using cladistic methods to work out how a fossil hominin taxon is related to modern humans and to other fossil hominin taxa.

The technical term *clade* refers to all (no more and no less) of the organisms descended from the most recent common ancestor. The smallest clade consists of just two taxa; the largest includes all living organisms. Cladistic analysis sorts taxa according to the amount of morphology they share, but the morphology has to be of a particular kind. To be helpful for working out relationships between closely related species, the morphology used must be shared by two or more taxa, but it must also vary within the group under investigation, so that it can be used to break that group up into subgroups, or clades. For example, the features that make all higher primates mammals, such as the presence of nipples and warm blood, are no use for sorting out detailed relationships among the great apes. But to go to the other extreme, morphology that is found only in *one* taxon cannot be used to work out the relationships *among* taxa.

Two taxa that share specialized morphology are referred to as "sister taxa." That pair of sister taxa has its own sister taxon (for example, *Gorilla* is the sister taxon of the *Pan/Homo* clade) and so on. The branching diagram that results is called a cladogram. The same relationships can be represented in writing by using sets of parentheses for sister groups (e.g., (((*Homo*, *Pan*) *Gorilla*) *Pongo*)).

Cladistic analysis works on the assumption that if members of two taxa share the same morphology, they must have inherited it from the same recent common ancestor. This assumption is often justified, but not always. We know that primates, including higher primates, have experienced convergent evolution, a process by which different lineages evolve similar morphology independently. The term *homoplasy* refers to similar morphology seen in two species but which is not inherited from a recent common ancestor. For example, it is likely that thick tooth enamel evolved more than once in human evolution, thus making it a homoplasy within the hominin clade.

Fossil DNA

The newest form of analysis used to work out how hominin taxa are related relies on the extraction and analysis of DNA. In your family, closely related individuals, for example, brothers and sisters, share more DNA than do distant cousins. It is the same for taxa. Individuals within a taxon should, on average, share more DNA than two individuals drawn from different taxa. However, despite the importance of DNA in our lives, fossilization quickly causes nucleic acids to degrade. For example, after fifty thousand years, only small amounts of DNA survive, and even this is broken into short fragments. A team led by Svante Pääbo, a molecular biologist from the Max Planck Institute for Evolutionary Anthropology at Leipzig, was the first to recover DNA from a fossil hominin, and I will consider fossil DNA evidence further when I discuss Neanderthals in Chapter 7.

Researchers undertaking fossil DNA analysis must take particular care to prevent and detect contamination. When people handle fossils, they inevitably leave hair and skin cells on the fossil, and these are a potent source of contamination. Scientists must make sure they are detecting DNA amplified from the fossil hominin and not DNA from other sources. In a recent study of cave bear fossil, researchers detected

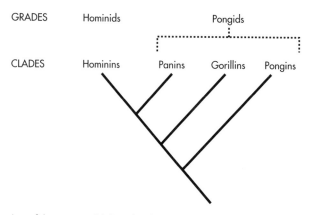

Comparison of the concepts of clades and grades as applied to living higher primates.

more than twenty different modern human DNA sequences on a single cave bear fossil. Tens, if not hundreds, of people, will have handled most hominin fossils, especially those found many years ago. Working out which of many DNA sequences recovered from a modern human fossil really belongs to that individual will be a challenge.

Grades

Homoplasy complicates our attempts to sort early hominins into clades. An alternative is to sort hominin taxa into grades. A grade is a category based on what an animal does rather than what its phylogenetic relationships are. So, for example, sport utility vehicles is the equivalent of a grade, whereas all the cars produced by the Ford Motor Company, including its range of SUVs, is the equivalent of a clade. Grades may also be clades, but they are not necessarily so. For example, leaf-eating, or folivorous, monkeys are a grade and not a clade because folivorous monkeys from the Old and New World are, respectively, just one component of much larger Old and New World monkey clades. A clade must comprise all the descendants of a common ancestor, not just some of them. Paleoanthropologists are more likely to agree about grades than clades, but determining the branching pattern of the TOL is something that must be pursued, even if the results are controversial. I will refer to some of these controversies in later chapters.

Functional and Behavioral Morphology

In addition to analyzing fossils in order to classify them and arrange them in a cladogram and then a phylogeny, paleoanthropologists also use the fossil record to work out the adaptations of hominin species. They do this by trying to reconstruct how individuals belonging to the same taxon lived their lives, and then they pool this information with evidence about habitat and generate hypotheses about how that species is adapted to its environment. Researchers try to learn as much about an extinct animal as they would expect to know about a living one. What did it eat? How did it move around? Did it live in social groups, or was it solitary? Paleoanthropologists attempt to answer these questions by looking at functional or behavioral morphology.

Functional morphology means looking at a bone or tooth and considering what functions it performs best and most frequently. For example, you would need curved finger bones only if you spent a lot of time holding on to branches, so curved finger bones are a sign that climbing was a part of that animal's locomotion. The shapes of finger joints and the length of the fingers and thumb also provide clues about how well early hominins could have gripped objects. Holding the shaft of a hammer needs a power grip, whereas the ability to hold and use a small, sharp stone tool uses a precision grip and a different combination of arm, forearm, and small hand muscles. Similarly, the thighbones of animals that bear all their weight on their hind limbs are differently shaped from those whose weight is distributed across all four limbs.

Functional morphology can also help to reconstruct the diet of early hominins. The shape of a tooth reflects what was eaten. Teeth with large crowns, with low, rounded, cusps covered by thick enamel are likely to have evolved to cope with a diet that included abrasive food, or food that was enclosed in some sort of hard outer coating, like the shell of a nut, that needed to be broken before the contents could be eaten. Scientists use microscopes to look at minute scratches not visible to the naked eye that are on all teeth. Foods like tubers that grow in the ground contain a lot of grit, and this leaves telltale gouges on the surface of the enamel. Sometimes teeth get scratched when animals trample them, or when hard sand grains are blown against them. But this type of damage should affect the sides and not just the top, or occlusal, surface of a tooth. When they look for clues about the diet of the early hominins by looking for evidence of any microscopic scratches left by food (called microwear), researchers must make sure that they do not confuse these scratches made after death (postmortem) with the scratches made during the life of the individual (antemortem microwear).

Direct evidence about the kinds of foods hominins ate comes from stable isotope analysis. This form of analysis measures oxygen, nitrogen, and carbon isotopes in fossil bones or teeth and then matches the pattern found in the fossil with the patterns seen in living animals whose diets are known. For example, animals that browse on leaves can be distinguished from those that graze on grass and from those that are primarily carnivores. Using such a method, Julia Lee-Thorp, an isotope chemist working at the University of Bradford's Department of Archaeological Sciences,

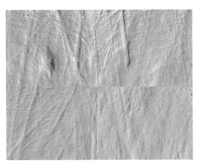

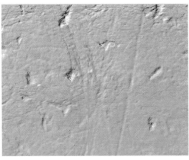

Photosimulation samples of antemortem dental microwear from seven specimens of *Paranthropus boisei* specimens known to preserve antemortem microwear. (From P. S. Ungar, F. E. Grine, and M. F. Teaford, "Dental Microwear and Diet of the Plio-Pleistocene Hominin *Paranthropus boisei*," *PLoS ONE* 3(4), 2008: e2044. doi:10.1371/journal.pone.0002044.)

and her colleagues have shown that 1.5 myr-old *Paranthropus* hominins from Swartkrans have stable isotope patterns that could come only from eating meat, thus causing researchers to reconsider earlier views that these hominins were primarily, if not exclusively, vegetarians.

Gaps and Biases in the Hominin Fossil Record

Over many decades paleoanthropologists have accumulated hominin fossils from thousands of individuals going back to between 6 and 7 mya.

While this number may sound impressive, the majority are concentrated in the later part of the hominin fossil record. Besides this temporal bias, the hominin fossil record has other biases and weaknesses. The science of working out these biases and trying to correct for them is the topic of "taphonomy." Whereas some of the hardest parts of the skeleton such as the teeth and the mandible are well represented in the hominin fossil record, the postcranial skeleton, that is, the vertebral column and the limbs, and particularly the vertebral column and the hands and feet, are poorly represented. The relative durability of different parts of the skeleton (for example, mandibles are generally heavier and are made of denser bone than vertebrae) is partly responsible for the differential preservation of body parts. Lighter bones like vertebrae are likely to be swept along in the floods that follow torrential rain, and then carried out into a lake, where they will be mixed in with the fossilized bones of fish and crocodiles. In contrast, heavier bones like skulls and jaws will fall to the bottom of the floodwaters, get trapped in the stones on the bed of the stream or river, and are thus preserved in sediments that preserve the heavier bones of other terrestrial animals.

Another factor influencing differential preservation is which parts of the carcass predators find most tempting. Leopards like to chew the hands and feet of monkeys, and if extinct large carnivores had similar preferences, then these parts of hominins would be in short supply as fossils. Thus, we know more about the evolution of the teeth of fossil hominins than we do about the evolution of their hands and feet. Body size also has a significant influence on whether a taxon is likely to have a fossil record. Large-bodied taxa are more likely to be fossilized than ones with small bodies, and larger individuals within a taxon have a greater likelihood of being fossilized than smaller members of the taxon. There is every reason to think that these same biases affect the hominin fossil record.

Some environments are more likely to lead to fossilization and subsequent discovery than others. Thus, we cannot assume that more fossil evidence from a particular period or place means that more individuals were present at that time, or in that place. It may just be that the circumstances at one period of time, or at one location, were more favorable for fossilization than they were at other times, or in other places. Likewise, the absence of hominin fossil evidence at a particular time or place does not have the same implication as its presence. As the saying goes, "absence of evidence is not evidence of absence." Similar logic suggests that taxa are likely to have arisen before they first appear in the fossil record, and they are likely to have survived beyond the time of their most recent appearance in the fossil record. Thus, the first appearance datum (or FAD) and the last appearance datum (or LAD) of taxa in the hominin fossil record are likely to be conservative statements about the times of origin and extinction of a taxon.

The same reservations apply to the geographical distribution of fossil sites. Hominins almost certainly lived in more locations than there are fossil sites. Environments in the past were often different than the ones we see now: parts of the world we now think of as being unattractive habitats were not necessarily that way in the past, and vice versa.

Lastly, not all environments are conducive to preserving bones and teeth. Some soils are so acidic that bones and teeth rarely survive. For a long time it was assumed that fossils would never be found in forested paleoenvironments because of the high levels of humic acid. This turned out to be a fallacy, but there are sites where archaeologists would have expected to see stone tools and bones together, and where they find only stone tools: the bones and teeth were dissolved before they could be fossilized.

FIVE

Possible Early Hominins

•

EIGHT MILLION YEARS AGO much of Africa was covered with thick forests interspersed with rivers and lakes, and most primates were tree dwellers. During the period from 8 to 5 mya the earth experienced the beginning of a long-term drying and cooling trend. The drying occurred because an increasing share of the earth's moisture was locked up in ice sheets that began to extend farther and farther away from the north and south poles. Temperatures fell, even in Africa, where the days were cooler and the nights cool, or even cold, at higher altitudes.

Hominin evolution began in Africa at the time of these climatic changes. Due to the increasing dryness, the dense forests were gradually replaced with open woodland. Tracts of grassland began to appear between large patches of trees. We tend to think that the grassland-adapted animals we associate with the modern-day African savannahs,

Acacia tree and grasslands in the Masai Mara Game Reserve, Kenya.

such as antelopes and zebra, have always been there. But they and the savannah they inhabit are relatively recent phenomena. The common ancestor of modern humans and living chimpanzees probably lived in the dense forest. Some of its descendants, though, began to adapt to life on the ground in more open conditions.

The fossil evidence for what may be the earliest hominins is found in sites which the other fossil and chemical evidence suggests were a mosaic of habitats—woodland, grassland, lakes, and gallery forests along rivers: no potential early hominin fossils have been found in an exclusively densely forested environment. This suggests that if these fossils do belong to early hominins, then the earliest hominins were adapted to both tree living and ground living. Trees would have provided fruit, nesting sites, and protection from predators. Patches of grassland would have provided new food sources such as tubers while lakes and rivers would have offered fish and molluscs. Although some early hominin fossils are found in caves it is unlikely that early hominins lived in the caves. Without a reliable source of heat and light, caves do not make attractive habitats for primates.

How to Tell an Early Hominin from an Early Panin

There are many differences between the skeletons of living chimps/bonobos and modern humans, particularly in the brain case, face, and base of the cranium, teeth, hand, pelvis, knee, and foot. There are also other important contrasts between the skeletons of modern humans and chimps/bonobos, such as the rates at which they develop and mature, and the relative lengths of the limbs, but you need better preserved skeletons than are usually seen in the early phases of the hominin fossil record to be able to detect these types of differences.

Table 3. Major Differences Between the Skeletons of a Modern Human and a Living Chimpanzee

	Modern Human	Chimpanzee
Forehead	Steep	Low
Face	Flat	Projecting
Cranial vault	Widest higher up	Widest at the base
Brain size	Large	Small
Canine teeth	Small	Large
Base of skull	Angled	Straighter
Thorax	Straight sides	Conical
Lumbar vertebrae	5	3–4
Limb bones	Straight	Curved
Limb proportions	Lower limb long	Lower limb short
Wrist	Less mobile	More mobile
Hand	Cup-shaped and long thumb	Flat, long fingers, and short thumb
Foot	Arched and big toe straight	Flat and big toe angled
Pelvis	Neonatal head is tight fit	Neonatal head has ++ room
Development— bones and teeth	Slow	Fast

However, all the differences listed in Table 3 are differences between the living members of the panin and hominin clades or lineages. Scientists searching in 8 to 5 myr-old sediments for the earliest hominins must consider a different question: what were the differences between the first hominins and the first panins? These are likely to have been much more subtle than the differences we see between contemporary hominins and panins. Although the *Pan/Homo* common ancestor was neither like a living chimp nor like a modern human, most researchers agree that it was probably more like a chimp than a modern human. The logic goes like this. Genetic and morphological evidence suggests that gorillas are the living animals most closely related to the chimp/bonobo and modern human common ancestor. Gorillas share more morphology with chimps than they do with modern humans (gorilla bones are more likely to be confused with the bones and teeth of a chimp than with the bones and teeth of a modern human). Therefore, the common ancestor of chimps and humans was probably more like a living chimp than a modern human, but "more like" does not mean "like." Its skeleton would most likely show evidence of being adapted for life in the trees. For example, its fingers would have been curved to enable it to grasp branches, and its limbs would have been adapted to walk both on all fours and on the hind limbs alone. Its face would have been snoutlike, not flat like that of modern humans, and its elongated jaws would have had relatively modest-sized chewing teeth, prominent canines, and large upper central incisor teeth.

The First Hominins

Researchers surmise that probably relatively little changed between the chimp/bonobo and modern human common ancestor and the earliest panins. But in what ways would the earliest hominins have differed from the chimp/bonobo and modern human common ancestor and from the earliest panins? Researchers predict that, unlike the earliest panins, it would have had smaller canine teeth, larger chewing teeth, and thicker lower jaws. There would also have been some changes in the skull and skeleton linked with more time spent upright and with a greater dependence on the hind limbs for bipedal walking. These changes would have included a forward shift of the foramen magnum, the place where the brain connects with the spinal cord, so that the head is better balanced on a body with a more vertical trunk, wider hips, straighter knees, and a more stable foot. But all this assumes there is no homoplasy, which is unlikely, especially among the members of the same adaptive radiation.

Simplicity Versus Complexity

Splitters and lumpers have very different models in mind for the early stages of hominin evolution. A lumper would entertain only three possibilities for an 8–5 myr-old higher primate fossil that was more closely related to modern humans and chimps/bonobos than to gorillas or orangs. It would either belong to the chimp/bonobo and modern human common ancestor, or be a primitive panin ancestral to living chimpanzees, or a primitive hominin ancestral to modern humans. Splitters who consider it likely that the first hominins and panins were just two of a number of closely related lineages would consider other options for the same 8–5 myr-old fossil. For them, in addition to the options listed above, it could belong to an extinct clade that is the sister taxon of the *Pan/Homo* clade, or to one, or more, extinct panin and hominin subclades.

Splitters would also expect to find evidence of homoplasies in this 8–5 mya period. Homoplasy complicates the task of sorting real hominins from taxa that may have independently evolved, and would thus share, one or more of the features researchers had assumed are only seen in hominins. Some researchers, and I am one of them, think we need

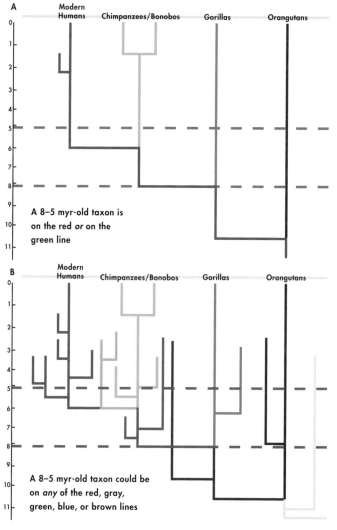

"Lumping/simple" (A) and **"splitting/complex"** (B) interpretations of the higher primate twig of the Tree of Life.

much better evidence than we presently have to be able to sort the earliest hominins from nonhominins with any degree of reliability.

Contenders for the Title of the Earliest Hominin

Researchers have put forward four species belonging to three genera as contenders for being the earliest hominin. One of the main problems in determining whether or not the fossils are actually primitive hominins is the small amount of fossil evidence we have for them. The fossil evidence for all four could fit comfortably in a supermarket trolley, and there would still be plenty of room to spare. Furthermore, the supermarket trolley does not necessarily contain the same evidence for each of the four contenders. Currently there is a distorted cranium, parts of several lower jaws and teeth of one, mostly teeth and some small hand and foot bones of a second, some teeth and parts of the thigh bones of a third, and a partial skeleton, several jaws and sets of teeth, of the fourth.

Sahelanthropus

The oldest of the contenders is *Sahelanthropus tchadensis*, known from hominin fossils discovered by Michel Brunet and his team from 2001 onward. It has been dated using relative biochronological methods to between 7 and 6 mya.

Sahelanthropus tchadensis is an important taxon for several reasons. First, it was found at a site called Toros-Menalla in Chad, in West Central Africa. This region is part of the Sahel, and just north of it today is the Sahara Desert. But 7–6 mya this region was very different. The geological and paleontological evidence suggests that the potential hominin lived in a complex habitat of lakes, grassy woodland, and rivers bordered by forests. We know this because geologists looking at

A recent photograph of the skull of *Sahelanthropus tchadensis*.

the rocks can identify traces of sediments that could only have been laid down on a lakeshore, and because vertebrates found at the site include freshwater fish and representatives of forest-dwelling, woodland, and grazing animals. Second, the hominin finds include a remarkably complete but distorted cranium as well as two mandibles. Researchers involved with interpreting the Chad finds have used virtual anthropology techniques to "straighten out" the cranium. This allows it to be compared more meaningfully with other later hominins and with chimpanzees.

The brain of the *S. tchadensis* cranium is chimp sized, but the upper part of its face has brow ridges like those seen in hominins less than half its geological age. The mandible is thicker than the jaws of living chimps, and the canines are worn down only at the tip and not also on the sides as they are in chimpanzees. Are the brow ridges, the robust lower jawbone, and the canines that wear down only at the tip sufficient evidence to be sure that *S. tchadensis* is a primitive hominin, and not the common ancestor of chimpanzees and humans, or a member of the panin lineage, or a member of another, extinct, clade?

Not all paleoanthropologists are convinced that *S. tchadensis* is a hominin. One view, almost certainly wrong, is that it is a fossil gorilla. If *S. tchadensis* is an early hominin, then the location of the site in West Central Africa means that the earliest hominins occupied a much wider area of Africa than paleoanthropologists previously thought.

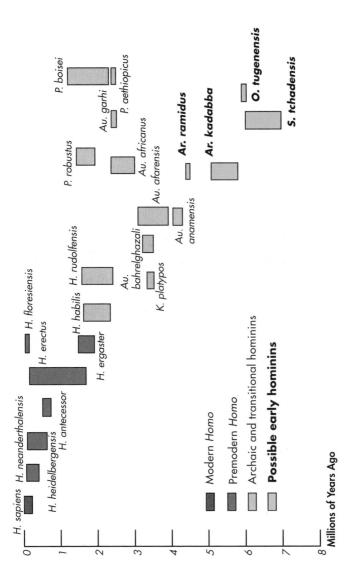

Time chart of "possible" early hominin species.

Orrorin

The second oldest potential primitive hominin species is *Orrorin tugenensis*, the name given to fossils found in sediments in the Tugen Hills of northern Kenya. Its age has been determined using potassium/argon dating to around 6 mya. One specimen, a lower molar tooth crown, was discovered in 1974, and twelve other specimens have been discovered since 2000.

The evidence for *O. tugenensis* is still frustratingly fragmentary. Its discoverers, Brigitte Senut and Martin Pickford, two paleoanthropologists based at the Collège de France in Paris, base their conclusion *O. tugenensis* is a hominin on two lines of evidence, one cranial, the other postcranial.

At a press conference in Nairobi, paleoanthropologists Martin Pickford and Brigitte Senut sit before the fossils they discovered in the Tugen Hills.

The cranial evidence relates to what Senut and Pickford claim is thick enamel covering the molar and premolar teeth of *O. tugenensis*. They suggest that enamel this thick is not found in panins and only in later, unambiguous, members of the hominin clade. But the researchers who found *O. tugenensis* place the greatest store on evidence from the part of the femur just below the hip joint. In climbing primates the outer, or cortical, bone is equally thick all round the neck of the femur, but in habitual bipeds the thickening is greatest at the top and bottom of the neck. Senut and Pickford claim that the cortical bone of the neck of the *O. tugenensis* femora is also preferentially thickened on the top and bottom of the neck. Unfortunately, their attempts to use CT to image the femoral neck have resulted in images that are so indistinct that it is not possible to be sure about the thickness of the bone around the neck.

Critics of the view that these fossils belong to an early hominin make three points. First, they say that the morphology of the *O. tugenensis* femur is not much different from that of primates that move around in trees. Second, it has not been demonstrated within higher primates that thick tooth enamel is confined to the hominin clade. Third, as Senut and Pickford admit, much of the morphology of the teeth of *O. tugenensis* is "apelike." Researchers have shown that the external shape of the *O. tugenensis* femur is like that of later archaic hominins. This strengthens the case for interpreting *O. tugenensis* as a hominin.

Until we have more evidence about *O. tugenensis*, it may be best to regard it as a creature closely related to the common ancestor of panins and hominins, but there is not enough evidence to be sure it is a hominin.

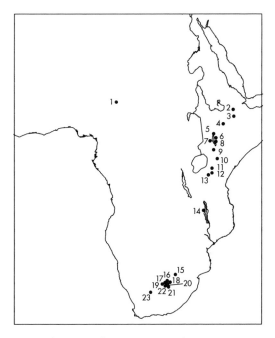

1 Koro Toro and Toros-Menalla	10 Lukeino *O. tugenensis*
Au. bahrelghazali, S. tchadensis	11 Peninj *P. boisei*
2 Hadar *Au. afarensis*	12 Olduvai Gorge *P. boisei*
3 Middle Awash/Gona *Au. afarensis*	13 Laetoli *Au. afarensis*
Ar. kadabba, Ar. ramidus,	14 Melema *P. boisei*
Au. garhi	15 Makapansgat *Au. africanus*
4 Konso *P. boisei*	16 Gondolin *P. robustus*
5 Omo *Au. afarensis, P. aethiopicus,*	17 Kromdraai *P. robustus*
P. boisei	18 Drimolen *P. robustus*
6 Koobi Fora *P. boisei ?Au. afarensis*	19 Sterkfontein *Au. africanus*
7 West Turkana *P. aethiopicus,*	20 Swartkrans *P. robustus*
P. boisei, K. platyops	21 Gladysvale *Au. africanus*
8 Allia Ba *Au. Anamensis*	22 Cooper's *P. robustus*
9 Kanapoi *Au. anamensis*	23 Taung *Au. africanus*

Map of Africa showing the main early and archaic hominin fossil sites. (Miller, Barbara D.; Wood, Bernard; Balkansky, Andrew; Mercader, Julio; Panger, Melissa, *Anthropology,* 1st, ©2006. Adapted and electronically reproduced by permission of Pearson Education, Inc., Upper Saddle River, New Jersey.)

Ardipithecus

The two other collections of fossils that might be from a primitive early hominin are both included in the same genus, *Ardipithecus*. The older fossil collection, dated to 5.7–5.2 mya, is assigned to *Ardipithecus kadabba* and comes from the Middle Awash region of Ethiopia. The fossils include mandibles, teeth, and some postcranial bones. Many aspects of the fossil evidence, such as the tall, pointed, upper canines, resemble chimpanzees. Little of the morphology of the fossils in this collection resembles that of the archaic hominins I discuss next. The case for regarding *Ar. kadabba* as a hominin is not a strong one.

The second collection of *Ardipithecus* fossils comes from the Middle Awash and Gona regions of Ethiopia. They date from around 4.4 mya and they may have persisted to around 4 mya. The fossil collection includes a partial skeleton, sets of teeth, part of the underside of a cranium, parts of several jaws, and some other limb bones. It is assigned to the genus *Ardipithecus*, but in a separate species called *Ardipithecus ramidus* because its discoverers think that its canines are less apelike than those of *Ar. kadabba*.

Several features link *Ar. ramidus* with hominins, the strongest evidence being the position of the foramen magnum and the relatively small (compared to chimps/bonobos) canines.

Information about the size of the brain of *Ar. ramidus* suggests that it is chimp/bonobo-like in size and shape, and evidence for its posture and locomotion is frustratingly contradictory. A very poorly preserved pelvis has been interpreted as being consistent with a bipedal locomotion, but the hands and feet are difficult, if not impossible, to reconcile with bipedalism. In terms of size, both *Ar. kadabba* and *Ar. ramidus* were similar to a small adult modern chimpanzee, around 70–80

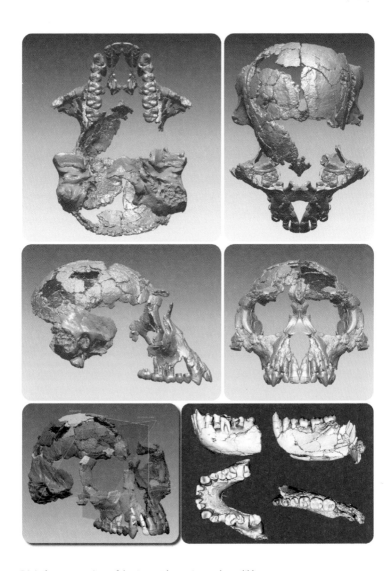

Digital representations of the *Ar. ramidus* cranium and mandible.

pounds. In spite of changes in the teeth and base of the skull in *Ar. ramidus* that link it with archaic hominins (discussed next), in overall appearance *Ar. ramidus* would have been much more like a chimpanzee than like a modern human.

Of the four potential hominins, three of them, *S. tchadensis*, *O. tugenensis*, and *Ar. ramidus* have different reasons for being included in the hominin clade. Whereas "splitters" would use the binomials I have used for the four taxa, "lumpers" would take the view that either all four taxa are different species within a single genus, *Ardipithecus*, or they all belong to a single species, *Ar. ramidus*.

Chimps Have Almost No Fossil Record

If modern humans and chimps/bonobos are each other's closest living relative, then both have been evolving separately for the same length of time. As we will see in the subsequent chapters of this book, modern humans have a substantial fossil record, much better than that for many other mammals. But the fossil record for the chimp/bonobo clade is virtually nonexistent. The only panin fossil evidence in the last 8 myr consisted of a few 700 kyr-old isolated teeth from a site called Baringo, in Kenya.

Odd? Certainly. In the past it has been "explained" that because chimps lived in the forest, and because there is little chance of erosion in the forest, then there are no exposures, and thus no places where fossils could be uncovered by erosion. Others say that high levels of humic acid in the soils of forests dissolve bones before they can be fossilized. Neither of these explanations is wholly convincing. Fossils are difficult to find in forests, but they are there. They just do not happen to include any fossil evidence belonging to panins. Of course, some of the fossils assigned to *Ardipithecus*, *Orrorin*, and to *Sahelanthropus* could be more closely related to panins than to hominins, but no one has been anxious to forgo the chance of being the discoverer of the earliest hominin in favor of being the discoverer of the earliest panin.

This is strange, because from the point of view of the wider biological interest it would be much more interesting to find fossil evidence of an early panin ancestor rather than fossil evidence of yet another early hominin. If we could find out what an early panin looked like, it would mean that researchers would have a better chance of identifying "real" hominins. There are other reasons why it would be helpful if researchers found an early panin. At the moment researchers make the assumption that the common ancestor of hominins and panins, and early panins, were more chimp/bonobo-like than modern human–like. It would be much better to know what early panins were like rather than having to make guesses about them, and this information would also help the researchers who are trying to identify homoplasies in the *Pan*/*Homo* clade.

·····

POINTS TO WATCH

- If the molecular evidence for the timing of the split that gave rise to the hominin and panin clades places it closer to 5 than 8 mya, then some possible early hominins like *S. tchadensis* may be ruled out because they antedate the split.

- When we have more fossil evidence from 5 to 8 mya, this should make it clearer whether the early phase of hominin evolution is "simple" or "complex."

- If researchers are able to locate rocks of the right age that sample more forested habitats, they may be able to locate more evidence of fossil chimpanzees and fossil evidence of gorillas.

· · · · ·

SIX

Archaic and Transitional Hominins

•

IN THIS CHAPTER I deal with creatures that are almost certainly hominins. They share substantially more of their morphology with modern humans than they do with chimpanzees. Yet they do not show the changes in jaw and tooth size and in body size and shape that characterize hominin species we include within our own genus *Homo*. So I call them "archaic" hominins. At the end of the chapter I also consider a group of hominins that seem to be part archaic hominin and part *Homo*: I call these "transitional" hominins.

Archaic Hominins from East Africa

Half a million years later in geological time than *Ar. ramidus*, between 3 and 4 mya, we begin to see signs of a creature with a much more

A **detail of a full-sized model** of Lucy is displayed at the Houston Museum of Natural Science in Houston, Texas, August, 29, 2007.

comprehensive fossil record than any of the potential primitive hominins discussed in the last chapter. The creature, an undoubted hominin, is called *Australopithecus afarensis*.

This was the name given in 1978 to fossils recovered from Laetoli in Tanzania and from the Ethiopian site of Hadar. The fossil record of *Au. afarensis* includes a skull, a partial skeleton, several well-preserved crania, many lower jaws, and sufficient limb bones to be able to generate reliable estimates of its size and body weight.

The Hadar part of the collection includes the famous "Lucy," which is close to half of the skeleton of an adult female individual. The find made by Don Johanson and his team made the headlines because it was the first time researchers had recovered such a well-preserved early hominin. Knowing the bones come from the same individual means that researchers can match jaws and teeth with limb bones, and arm bones with leg bones. It also means they can make more accurate estimates of stature, body weight, and the relative length of the limbs.

The picture of *Au. afarensis* that emerges is of a hominin weighing from 75 pounds to 125 pounds (34 to 57 kilograms). Its brain volume was between 400 and 500 cm³,

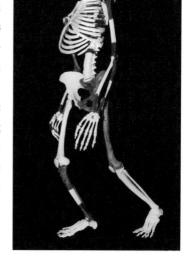

Reconstructed skeleton of Lucy at the National Museum of Ethiopia, in Addis Ababa.

larger than the average brain size of a chimpanzee and substantially larger than the 300–350 cm³ estimate for the brain size of *S. tchadensis* and *Ar. ramidus*. However, when brain size is related to the size of the body (blue whales have larger brains than modern humans, but they weigh more than we do) the brain of *Au. afarensis* is only a little larger than that of an equivalent-sized chimpanzee. Its incisor teeth (the four teeth in each jaw you see when people smile) are much smaller than those of chimps, but the chewing teeth (the two premolars and three molars on each side that are at the back of the jaw—you need to make someone laugh out loud to see them) of *Au. afarensis* are larger than those of chimps. This suggests that its diet included more hard-to-chew items than does the diet of chimps. The shape and size of the pelvis and lower limb remains suggest that *Au. afarensis* was capable of walking bipedally but probably only for short distances.

The oldest preserved trails of hominin footprints, and the oldest hominin trace fossils, are the 3.6 myr-old trails excavated at Laetoli, Tanzania, by Mary Leakey. The hominin footprints are just one of many trails made by large and small animals, ranging in size from horses to hares. The foot- and hoofprints are well preserved because the animals happened to walk across a flat area where a layer of volcanic ash had recently been moistened by a rainstorm. The type of fine volcanic ash at Laetoli has a chemical content that makes it behave like cement, so when the sun dried out the layer it became rock hard. The process is much like the one used outside a Hollywood restaurant to preserve the hand- and footprints of film stars. These trace fossils provide graphic evidence that a contemporary early hominin, presumably *Au. afarensis*, was capable of walking bipedally. The size of the footprints and the length of the stride are consistent with estimates of stature made using the limb bones of *Au.*

A polyester cast of three sets of 3.6 million-year-old hominid footprints at Laetoli in Tanzania, August 16, 1996.

afarensis, suggesting that the standing height of individuals was between 3 and 4 feet.

Fossils from a site in Kenya called Kanapoi that date from 3.9–4.2 mya belong to a different hominin, *Australopithecus anamensis*, that might be ancestral to *Au. afarensis*. The canines of *Au. anamensis* are more chimplike than those of *Au. afarensis*, yet the chewing teeth are very different from those of chimps. Three-and-a-half-million-year-old hominin fossils collected at Bahr el ghazal in Chad in 1995, not far from the site where *S. tchadensis* was found subsequently, have been assigned to *Australopithecus bahrelghazali*, but some researchers claim, probably correctly, that these remains do not belong to a separate hominin species, but to a geographical variant of *Au. afarensis.*

The fourth East African archaic hominin, the 2.5 myr-old *Australopithecus garhi* found at Bouri, in the Middle Awash of Ethiopia, is in many ways the strangest. Limb bones found with it suggest it was a biped, but its chewing teeth are a good deal larger than those of the other three East African australopiths. No stone tools have been found with the *Au. garhi* fossils, but animal bones found close by show telltale signs that flesh had been removed using a sharp-edged tool. Only razor-sharp stone flakes wielded by a hominin would have allowed the flesh to be removed so neatly. This is currently the oldest evidence that hominins were deliberately defleshing animal carcasses.

Archaic Hominins from Southern Africa

All the australopith taxa I have introduced thus far have been found in East or Central Africa at sites on the open landscape. The localities where the hominin fossils were found were not necessarily places where the hominins lived or camped: they were simply places on the landscape where, for one natural reason or another, one or more hominin bones had accumulated. Maybe they were transported there by the runoff from a rainstorm, or the site may have been close to the food cache or lair of a predator. Most of the sites have been dated by applying isotope-dating methods to volcanic ash either in the same horizon as the hominin fossil evidence is likely to have come from, or in layers above and below the fossil-rich layer.

However, in 1924, nearly fifty years before the discovery of the remains belonging to *Au. afarensis*, the skull of a hominin child was discovered in southern Africa in a very different context. It was discovered among the fragments of bone that came from a small cave exposed during mining at the Buxton Limeworks at Taung. The new hominin was drawn to the attention of Professor Raymond Dart, who was the first expert to recognize its significance.

Dart called the new taxon *Australopithecus africanus*, which means literally the "southern ape of Africa." When he wrote about the new find in an article in *Nature* in 1925, he received a frosty reception. Most researchers were either ignorant of, or had forgotten, Darwin's prediction about Africa being the origin of humankind. However, Dart managed to recruit a distinguished ally, the paleontologist Robert Broom, who had made a name for himself by collecting fossils of mammal-like reptiles. Broom was so convinced that Dart had found an important link between our ape ancestors and modern humans that he started to look for other caves that might contain the bones of *Au. africanus*, or of creatures like it.

Broom searched for more than a decade before a second hominin-bearing cave site, Sterkfontein, was discovered. It contained remains that scientists now interpret as belonging to the same species as the Taung child. Soon after came discoveries at two more caves, Kromdraai and Swartkrans, of creatures whose chewing teeth, faces, and jaws differed from those of *Au. africanus*. These remains were allocated to a different genus and species, *Paranthropus* (which means "beside Man") *robustus*. Its slightly larger chewing teeth just about put it into the "megadont (meaning large-toothed) archaic hominin" category. More recently hominin fossils have been found at other southern African cave sites (e.g., Drimolen and Gladysvale), but all these recent finds seem to belong to either *Au. africanus* or *P. robustus*.

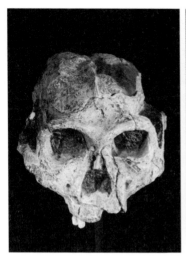

Australopithecus africanus (left) and *Paranthropus robustus* (right) on display at the Anthropological Institute and Museum of the University of Zurich.

Interpreting the Southern African Hominins

One problem with interpreting the hominins recovered from the southern African caves is that they cannot be dated as reliably as those from sites in East Africa. At all these southern African cave sites early hominin fossils are mixed in with other animal bones in hardened rock and bone-laden cave fillings, or breccias. Researchers are trying to find absolute dating methods that will work on the cave breccias, but in the meantime most of these sites have only been dated by comparing the remains of the mammals found in the caves with fossils found at the better-dated sites in East Africa. In this way the ages of the *Au. africanus*–bearing breccias are estimated to be between 2.4 and 3 mya. A remarkably complete hominin skeleton, numbered Stw 573, from deep in the Sterkfontein cave, may be older, but it is too early to tell whether it belongs to *Au. africanus*. Hominins resembling *Au. africanus* recovered from even deeper in the Sterkfontein cave system, from the Jacovec Cavern, may also be older than those found in the main cave.

Our current understanding of *Au. africanus* is that its physique was much like that of *Au. afarensis*, but its chewing teeth were larger, its skull was not as apelike, and its limb proportions were more apelike. Its average brain volume is a little larger than that of *Au. afarensis*. The postcranial skeleton suggests that, although *Au. africanus* could walk bipedally, it was also capable of climbing in trees. The other animal fossils and the plant remains found with *Au. africanus* suggest that its habitat was grassy woodland. The picture we have of the 1.5–2 myr-old *Paranthropus* differs in that its chewing teeth are larger, its face is broader, and its brain is slightly bigger. Some researchers think that the locomotion of *P. robustus* may have differed from that of *Au. africanus*, but there is not enough evidence to be sure of this.

There is no sign that either *Au. africanus* or *P. robustus* lived in the caves. Either their bones were dropped into cave openings by leopards, or they were brought into the caves by hyenas or porcupines. Some of the more complete remains like that of the Stw 573 skeleton from Sterkfontein may belong to individuals who had either fallen into the caves or who had explored them and found them easier to enter than to leave.

Really Megadont Archaic Hominins in East Africa

Further evidence that *Paranthropus* was distinct from *Au. africanus* came in 1959 when Mary and Louis Leakey discovered a 1.9 myr-old fragmented cranium at Olduvai Gorge in Tanzania. The OH 5 cranium has much larger chewing teeth and jaws than *P. robustus*, but its incisors and

Mary and Louis Leakey study fossilized skull fragments in Tanganyika, 1959.

canines are small, both absolutely and in relation to the size of its premolars and molars. Whatever these creatures were eating, they evidently did not need large incisors to bite into it.

The OH 5 cranium was made the type specimen of *Zinjanthropus boisei*, but most researchers have dropped the genus *Zinjanthropus* and place this East African taxon into either *Australopithecus* or *Paranthropus*: I will refer to it as *Paranthropus boisei*. Further evidence of *P. boisei* came with the discovery of a mandible with a large, robust body, large chewing teeth, and small incisors and canines at the Peninj River, on the shores of Lake Natron, in Tanzania. Since then more fossils belonging to *P. boisei* have been found at Olduvai, and at sites in Ethiopia, Kenya, and Malawi.

The features that set *P. boisei* apart are found in the cranium, the mandible, and the dentition. It is the only hominin to combine a massive, wide, flat face with very large chewing teeth and small incisors and canines. Despite these large jaws and chewing teeth, its brain (around

450 cm³) is similar in size to the brains of australopiths like *Au. africanus*. The earliest evidence of *Paranthropus* in East Africa is a variant that has a more projecting face, larger incisors, and a more apelike cranial base. Some researchers assign these pre-2.3 mya fossils to a separate species, *P. aethiopicus*.

Despite the richness of the cranial evidence for *P. boisei*, no postcranial remains have been found in association with cranial remains that we can be sure belong to *P. boisei*. So, we have no good evidence, only guesswork, about its posture or locomotion.

Most paleoanthropologists interpret the large-crowned, thick-enameled chewing teeth, the large mandibles with wide bodies, and the crests on the crania of large individuals as evidence that the diet of *P. boisei* was highly specialized, perhaps consisting mainly of seeds, or fruits with hard outer coverings. Others disagree and say that *Paranthropus* may have been the higher primate equivalent of a bush pig. Its large chewing teeth and mandibles would have enabled it to cope with a wide range of dietary items including plant foods, insects, and molluscs. But paradoxically jaws and teeth may not reflect what animals are eating *most* of the time, but what they are driven to eat when their preferred foods are not available. Being able to access these so-called fallback foods may prevent an animal from starving during periods of drought. Whatever *P. boisei* ate either some, or most, of the time it was certainly abrasive, for in most individuals even their excessively thick enamel is nearly worn away in ways that you rarely see in modern humans.

There are enough skulls and crania to see that *P. boisei* showed a modest increase in brain size through time. From what little we know of their hands there is no morphological reason why *P. boisei* or *P. robustus* could not have made primitive stone tools. Pointed sticks found with

A pencil drawing of *Paranthropus boisei.*

P. robustus show wear that matches that produced by contemporary hunters and gatherers when they use sticks to break into termite hills for the energy-rich and palatable termites.

The largest specimens of *P. boisei*, almost certainly males, were almost twice the weight of the smallest, presumably female, individuals (around 150 pounds [68 kilograms] compared to 75 pounds [34 kilograms]). In living primates such a wide range of body size is associated with a social system in which there is competition among males for access to females. In comparable living primates, males establish this hierarchy through threats mediated by displaying their large canine teeth. The absence of large canines in *Paranthropus* suggests that if there was a male dominance hierarchy, then male *Paranthropus* individuals must have used some other means for establishing it. Perhaps the sheer size of their faces, combined perhaps with orangutan-like skin folds, could have been the means they used to establish their place in the hierarchy.

Kenyanthropus

The latest archaic hominin to be discovered was assigned to a new genus and species called *Kenyanthropus platyops*. This is the name that in 2001 Meave Leakey and her colleagues gave to a collection of fossils recovered from horizons that are absolutely dated to between 3.3 and 3.5 mya. The best specimen is a cranium, but it is deformed by many matrix-filled cracks that permeate the face and rest of the cranium. Despite the cracking there are features of the face that do not match the face of *Au. afarensis*, the hominin best known in this time period. Meave Leakey's team is convinced their find is distinct from *Au. afarensis*, and they also point to the similarities between it and a taxon I will discuss in the next section, *Homo rudolfensis*. However, at this stage in their investigation they are unsure whether the facial similarities are inherited from a recent common ancestor (an apomorphy) or whether the shared facial morphology arose independently in the two taxa (a homoplasy).

Transitional Hominins

In 1960 at Olduvai Gorge, near where they had recovered the *P. boisei* cranium in 1959, Louis and Mary Leakey made the first of a series of remarkable discoveries of what they thought was a much more human-like early hominin than the archaic hominins I have considered up until now. Even today scientists are debating whether these remains belong to

Olduvai Gorge, in Tanzania, the site of many of Louis and Mary Leakey's hominin discoveries, is about thirty miles (forty-eight kilometers) long.

a primitive species of our own genus *Homo*, or whether they belong to a larger-brained archaic hominin.

The first finds consisted of some teeth, part of the top of a cranium, some hand bones, and most of a left foot. The next year the Leakeys found the incomplete skull of an adolescent, more cranial fragments, a lower jaw, and teeth. The cranial remains showed no sign of the bony crests characteristic of large-bodied *P. boisei* individuals, and the premolar and molar teeth were substantially smaller than those of *P. boisei*. Although the brain was small, Louis Leakey and Phillip Tobias, a distinguished South African anatomist from the University of the Witwatersrand initially recruited by the Leakeys to describe their 1959 *Zinjanthropus* cranium, were convinced that impressions on the inside of the cranial cavity provided evidence of Broca's area, the part of the brain that scientists at the time believed was the sole control center for the muscles involved in speech.

Louis Leakey, Phillip Tobias, and fellow anatomist John Napier argued that the material justified establishing a new species, *Homo habilis*, literally "handy man," within the genus *Homo*. Prior to their suggestion the consensus was that all *Homo* species should have a brain size of at least 750 cm³. The brains of the new Olduvai discoveries, however, were only about 600–700 cm³. Louis Leakey and his colleagues argued that the Olduvai evidence for *H. habilis* satisfied the functional criteria for *Homo*, namely, dexterity (for by now they were convinced that *H. habilis* and not *P. boisei* had made the stone tools that had been found in the same levels at Olduvai), upright posture, and a fully bipedal locomotion.

Similar fossils have since been recovered from other sites in East and southern Africa, but the single largest addition to the collection has come from the site of Koobi Fora in Kenya. The brain size of the enlarged sample of *H. habilis* ranges from just less than 500 cm³ to about 800 cm³. Some of the faces are small and projecting and others are large and flatter. The lower jaws also vary in size and shape. The limb bones found with *H. habilis* cranial remains show that its skeleton was like that of the archaic hominins in that it had long arms relative to the length of its legs. There is sufficient fossil evidence to generate an estimate of its limb proportions, and they are indistinguishable from those of *Au. afarensis*.

Taking all the new evidence into account, there is little to distinguish *H. habilis* from the archaic hominins. When we relate the size of its jaw and teeth to estimates of its body size, *H. habilis* is more similar to the australopiths than to later *Homo*. The conclusion that *H. habilis* was capable of spoken language was based on presumed links between Broca's area in the brain and language production that are no longer valid: we

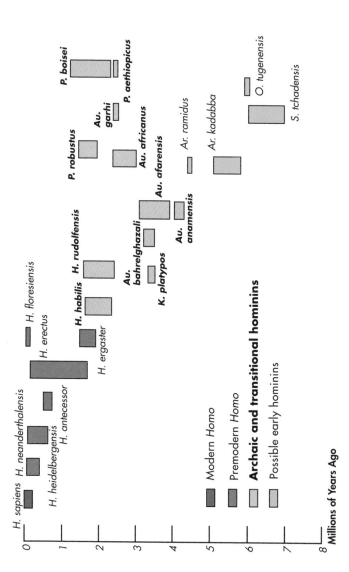

Time chart of "archaic" and "transitional" hominin species.

H. sapiens *H. neanderthalensis*
H. heidelbergensis
H. antecessor
H. floresiensis
H. erectus
H. ergaster
H. habilis *H. rudolfensis*
Au. bahrelghazali
K. platyops
Au. anamensis
Au. afarensis
Au. africanus
Au. garhi
P. robustus
P. boisei
P. aethiopicus
Ar. ramidus
Ar. kadabba
O. tugenensis
S. tchadensis

Modern *Homo*
Premodern *Homo*
Archaic and transitional hominins
Possible early hominins

Millions of Years Ago

0
1
2
3
4
5
6
7
8

now know that language function is more widely distributed across the brain. The postcranial skeleton of *H. habilis* differs very little from that of *Australopithecus* and *Paranthropus*. The hand bones found at Olduvai suggest *H. habilis* was capable of the manual dexterity involved in the manufacture and use of simple stone tools, but this is also true of the hand bones of *Au. afarensis* and *P. robustus*.

Researchers also generally agree that the crania, jaws, and teeth of *H. habilis* are more variable than one would expect for a single species. Many, but not all, researchers now divide it up into two species: *H. habilis* proper (technically called *sensu stricto*, i.e., in the "strict sense") and *Homo rudolfensis*. Compared to *H. habilis* proper, the latter has a bigger brain (700–800 cm^3), a bigger, wider, flatter face, and larger chewing teeth, suggesting that its diet may have differed from that of *H. habilis*. We know nothing for certain about the limbs of *H. rudolfensis*.

· · · · ·

POINTS TO WATCH

- Additional fossil evidence for *Au. anamensis* and *Au. afarensis* may well demonstrate that they, along with *P. aethiopicus* and *P. boisei*, are examples of new species forming by a speciation process called anagenesis.

- The jury is still out about whether the megadont hominins found in East and southern Africa are more closely related to each other than to any other extinct hominin. This will be resolved either by new fossil evidence, or by finding new ways to use the existing evidence, to demonstrate that the features found in all *Paranthropus* taxa are unlikely to be homoplasies.

- The case for keeping the two transitional hominin taxa, *H. habilis* and *H. rudolfensis*, within *Homo* would be greatly strengthened if the limb bones of *H. rudolfensis* were like those of *H. ergaster*. This needs the discovery and recovery of an associated skeleton of *H. rudolfensis*.

- Researchers are using evidence from morphological, functional, and isotopic studies to reconstruct the diet of *Paranthropus* species in order to determine whether their derived morphology (especially that of *P. boisei*) evolved as a response to the need to focus on a few food items as "fallback" foods, or as a way of coping with many different sorts of foods.

- Researchers would like to know what sorts of stone tools were made by archaic hominins. This may be difficult because the early stages of tool making may have been at a very low frequency, perhaps too low to show up as a conventional archaeological site.

• • • • •

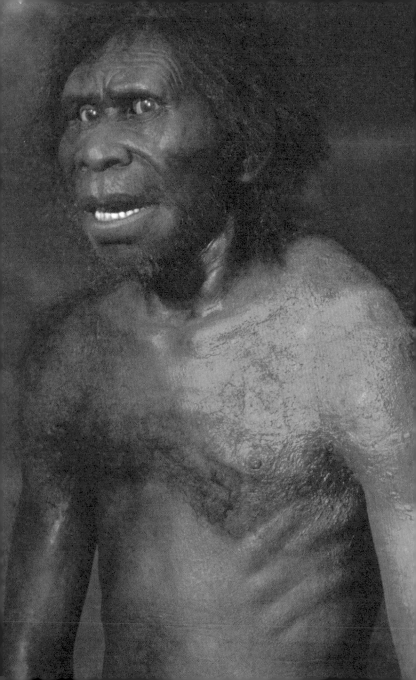

SEVEN

Premodern *Homo*

•

ALL THE TAXA that can confidently be considered as hominins, and I have considered thus far, are relatively small (ca. 60–120 lb [27–54 kg]) compared to most modern humans. Brain size and limb proportions are known for only a few individuals belonging to archaic and transitional hominin taxa. In all cases where there is enough information to make even a rough estimate of brain size, the brains are all below the absolute and relative size of later *Homo* taxa, and all of the taxa have relatively shorter legs than modern humans. This would have made them less efficient bipeds than we are, but it does mean that they would still have been able to climb into trees for shelter and to feed. The large chewing teeth and thick mandibular bodies of the archaic and transitional hominins, and the very large mandibular bodies and chewing teeth of the megadont archaic hominins, suggest that their diet routinely or occasionally

Reconstruction of *Homo erectus*.

included tougher or more abrasive food than the diets of modern humans. All the archaic hominins and the transitional hominins seem to belong to a different grade than modern humans. So when and where in human evolutionary history do we see the earliest evidence of creatures that are more like modern humans?

Homo ergaster

A little less than 2 mya we begin to see in some of the fossils recovered from Koobi Fora and West Turkana, both sites in Northern Kenya, the first evidence of creatures that are more like modern humans than any archaic or transitional hominin. The formal name for this fossil evidence is *Homo ergaster*, but most researchers do not use a separate species name for this material, and instead refer to it as belonging to "early African *Homo erectus*," or just *Homo erectus*.

Homo ergaster is the first hominin with a body whose size and shape is more like that of modern humans than any of the archaic or transitional hominin taxa. In relation to

Reconstruction of the *Homo ergaster* individual known as Turkana Boy, whose skeleton was found at West Turkana in Kenya in 1984.

the size of its body, the teeth and jaws of *H. ergaster* are smaller than those of the archaic and transitional hominins. This means *H. ergaster* either had a different diet than that of the archaic and transitional hominins, or it was eating the same sorts of food, but was processing them outside the mouth instead of inside the mouth. The obvious way to process food outside the mouth is to cook it, and several researchers have suggested that *H. ergaster* may have been the first hominin to routinely cook food. Cooking makes some tough foods easier to eat, and it also renders inactive many of the chemicals that make otherwise nutritious food poisonous.

The earliest evidence of burned earth close to where stone tools have been found is dated to between 1 and 2 mya. It is tempting to interpret this as evidence of deliberate fire, but when lightning strikes a tree and sets it on fire, the remains of a burned tree stump can be confused with the remains of a controlled fire made in a hearth. Controlled fires usually burn hotter than natural fires in tree stumps, but while in theory it should be possible to tell the remains of a natural fire from a hominin-controlled fire it is not always so easy. The earliest archaeological evidence of the ability to control fire presently comes from the ca. 800 kyr-old site of Gesher Benot Ya'aqov in Israel: evidence of stone hearths does not come until much later (ca. 300 kya) in the archaeological record.

The long lower limbs of *H. ergaster* would have allowed them to travel long distances efficiently. Clearly some adult modern humans are adept at climbing trees to recover nuts and honey, but modern humans are not adapted to travel any significant distance in the trees. Their long legs get in the way, and their arms have lost the apelike ability to use branches efficiently for locomotion. In all these aspects *H. ergaster* is more specialized in the direction of modern humans than earlier hominins. However, in one important respect, brain size, *H. ergaster* shows little advance over

H. rudolfensis, the larger brained of the two transitional hominin taxa. Why large brains do not appear until much later in human evolution is still a puzzle to paleoanthropologists. Perhaps it may have been related to the avoidance of the extra risk in the later stages of pregnancy. The shape and size of the true pelvis, combined with what can be extrapolated from adult brain sizes about the brain size of a *H. ergaster* neonate, suggests that the head was small enough to be oriented transversely all the way through the birth canal, and thus it did not need to be rotated after negotiating the pelvic inlet. This would have effectively eliminated one of the common causes of obstructed labor in modern humans.

Out of Africa: Who and When?

Until just less than 2 mya the hominin fossil and the archaeological records are confined to Africa. But "absence of evidence is not evidence of absence" so we must be aware of falling into the trap of ceasing to look for evidence of hominins outside Africa before this time.

Currently the earliest good fossil evidence of hominins beyond Africa comes from the site of Dmanisi in the Caucasus. There are no absolute dates for the sediments from the site, but the radioisotope age of the lava beneath the sediments and the fossil animals found with the hominins suggest an age of around 1.7–1.8 mya. The hominins found there have yet to be studied in detail, but they appear to belong to a relatively primitive *H. ergaster*-like creature. However, what is intriguing is that the stone tools recovered from the same horizon as the Dmanisi hominins are like the earliest African stone tools that archaeologists refer to as belonging to the Oldowan (they are named after Olduvai Gorge, Tanzania, the site where they were first found) Culture. After Dmanisi, the next oldest well-dated evidence of hominin occupation in the region is the 1.5 myr-old site of 'Ubeidiya in Israel, but so far only a few hominin teeth have been found there.

Homo erectus

By one million years ago evidence of a new type of hominin, *Homo erectus*, is found in Africa, China, and Indonesia. Some researchers, but not all, are persuaded that *H. erectus* first reached Indonesia as early as 1.7 mya, and perhaps as early as 1.9 mya. If so, they would most likely have been established on the Asian mainland sometime before that. At present stone tools dated to 1.5 mya are the earliest reliable evidence of hominins in what is now modern-day China.

If you met a *H. erectus* in the street, you would be unlikely to confuse it with a modern human, but it would be much more like a modern human than any archaic or transitional hominin. The best-known fossil evidence of *H. erectus* comes from sites along the Solo River on the island of Java, Indonesia, and from the Peking Man site (now called Zhoukoudian) in China. As we saw in Chapter 3, Eugène Dubois found the first *H. erectus* fossils in Java. Encouraged by finding a small piece of lower jaw at a site called Kedung Brubus in northern Java, Dubois turned his attention to one of the parts of Java where the Solo River has exposed sediments that we now know may date back to around 2 mya. He organized an elaborate excavation of the sediments that are exposed in the banks of the river during the dry season near the village of Trinil. In 1891 the excavators uncovered some teeth, a femur, and a skullcap (technically this is called a calotte). Initially he thought the calotte belonged to an extinct ape, but he evidently changed his mind because in 1894, two years after the initial publication, he published a paper giving it a different genus name, *Pithecanthropus*. Researchers now include *Pithecanthropus* in the

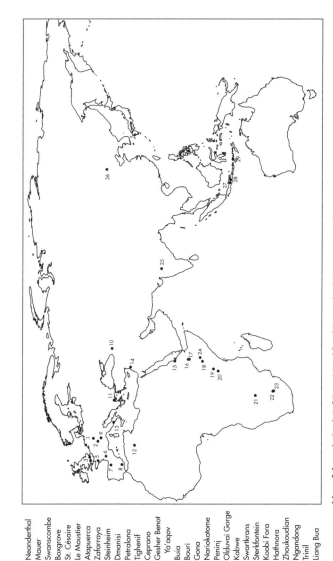

1 Neanderthal
2 Mauer
3 Swanscombe
4 Boxgrove
5 St. Césaire
6 Le Moustier
7 Atapuerca
8 Zafarraya
9 Steinheim
10 Dmanisi
11 Petralona
12 Tighenif
13 Ceprano
14 Gesher Benot
   Ya'aqov
15 Buia
16 Bouri
17 Gona
18 Nariokotome
19 Peninj
20 Olduvai Gorge
21 Kabwe
22 Swartkrans
23 Sterkfontein
24 Koobi Fora
25 Hathnora
26 Zhoukoudian
27 Ngandong
28 Trinil
29 Liang Bua

Map of the main "archaic," "transitional," and "premodern" *Homo* sites. (Miller, Barbara D.; Wood, Bernard; Balkansky, Andrew; Mercader, Julio; Panger, Melissa, *Anthropology*, 1st, ©2006. Adapted and electronically reproduced by permission of Pearson Education, Inc., Upper Saddle River, New Jersey.)

genus *Homo*. Remember that in 1894 the only two hominin taxa known were modern humans, *Homo sapiens*, and the Neanderthals, *Homo neanderthalensis*. The Trinil specimen lacks the large brain and tall rounded brain case of modern humans. Its brain volume was about 60 percent of the average for modern humans, but the femur found close by looked like a modern human femur, and this is why Dubois called his new species *Pithecanthropus erectus*. However, not all researchers are convinced that the femur is as old as the calotte. It may belong to a much more recent skeleton, and may have been "reburied" in the river gravels. The search for hominins at Trinil continued for a decade; the last hominin fragment to be recovered from the site was found in 1900.

The focus for the next phase of the search for hominin remains in Java was upstream of Trinil, where the Solo River cuts through the sediments of the Sangiran Dome. It was here that in 1936 a German paleontologist, Ralph von Koenigswald, recovered a cranium that resembled the Trinil skullcap, but the brain size was even smaller than that of the Trinil calotte. Several more specimens were recovered, but then the Second World War and the Japanese occupation of Java curtailed research. Ralph von Koenigswald temporarily buried the hominin fossils in gardens in order to hide them from the Japanese. The search for early hominins was renewed after the Second World War, and research in and around the Sangiran

In this undated photograph, German paleontologist Ralph von Koenigswald is shown examining the upper jaw of *Pithecanthropus robustus.*

Dome is ongoing. Researchers have recovered mandibles, several crania, and some postcranial evidence.

Whereas there was a lull in research activity in Java in the 1920s, in China the early 1920s marked the beginning of the search for early hominins. A Swedish paleontologist, Gunnar Andersson, and a junior colleague from Austria, Otto Zdansky, excavated for two seasons, in 1921 and 1923, at the Zhoukoudian (formerly spelt Choukou-tien) Cave, near Beijing. They recovered quartz artifacts, but apparently there were no fossil hominins. However, in 1926, when he was reviewing the excavated material shipped to Uppsala, Zdansky realized that two of what had been labelled as "ape" teeth from Locality 1 belonged to a hominin. The teeth, an upper molar and a lower pre-molar, were described by the anatomist Davidson Black in 1926, and together with a well-preserved left permanent first lower molar tooth found in 1927, they were assigned to a new genus and species *Sinanthropus pekinensis* by Black.

In the same year Black, together with Anders Bohlin and a Chinese colleague, Weng Wanhao, resumed excavations at Zhoukoudian. The first cranium was found in 1929 and excavations continued until they were interrupted by the Second World War. The fossils recovered from Locality 1 were all lost during the war. They were to be shipped to the United States, but they never arrived. Their whereabouts remains a mystery. They were apparently to be taken to a place of safety by a unit of US Marines. It is not clear whether the fossils were lost before the marines reached a port, or whether they were lost at sea. Even today people come forward claiming a relative has bequeathed them a trunk full of priceless early hominin fossils. Luckily excellent casts had been made at the American Museum of Natural History, and one

A depiction of *Homo erectus* in the cave at Zhoukoudian made before the evidence for controlled fire was shown to be problematic.

of the AMNH scientists, Franz Weidenreich, had prepared meticulous qualitative and quantitative descriptions of the material. Some of its morphology was distinctive, yet in many other ways the *Sinanthropus* fossils resembled those belonging to *Pithecanthropus erectus* from Java. In order to recognize this, in 1940 Franz Weidenreich suggested that both sets of fossils should be merged in a single species called *Homo erectus*. Since the Second World War fossils similar to those belonging to *Pithecanthropus* and *Sinanthropus* have been found at other sites in Java (e.g., Ngawi and Sambungmacan), China (e.g., Lantian), and southern (e.g., Swartkrans) and East (e.g., Melka Kunturé, Middle Awash, Olduvai Gorge, and Buia) Africa.

Despite the recovery of a relatively large number of crania from Java, China, and elsewhere in the last century, relatively little was known about the limbs of *H. erectus*. This situation changed with the discovery in East Africa of crucial postcranial evidence. This came in the form of a pelvis and femur from Olduvai Gorge (OH 28), two fragmentary partial skeletons from Koobi Fora (KNM-ER 803 and 1800), and an unusually well-preserved skeleton from West Turkana (KNM-WT 15000).

If the antiquity for the child's cranium from Modjokerto/Perning, and the very recent date for the Ngandong remains are confirmed, then,

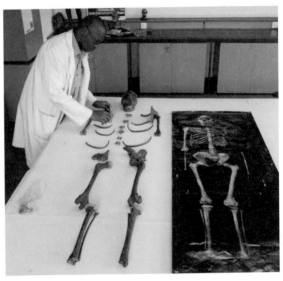

This photograph, taken on February 1, 2007, shows Samuel Muteti, at the Kenya National Museum in Nairobi, displaying a replica of the Turkana Boy.

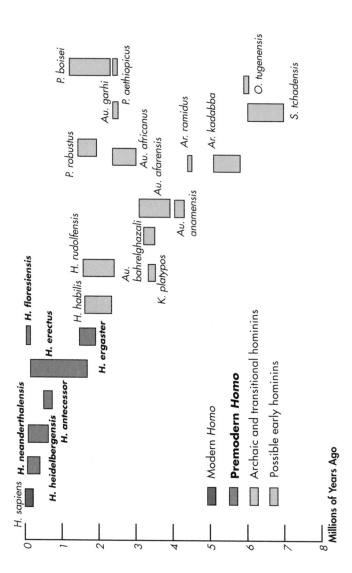

Time chart of "premodern" *Homo* species.

even if *H. ergaster* from East Africa is excluded from the *H. erectus* hypodigm, the two sets of dates suggest the temporal range of *H. erectus* was from ca. 1.9 mya to ca. 50 kya. The crania of *H. erectus* are all low, with the greatest width low down on the cranium. There is a substantial and more or less continuous bony ridge, or torus, above the orbits, a depression, or sulcus, behind it, and a pronounced blunt ridge (or keel) of bone runs in the midline from the front to the back of the brain case: this is called a sagittal torus. At the back of the cranium the sharply angulated occipital region has a well-defined sulcus above it. The walls of the brain case are made of two layers, or laminae, of bone. In *H. erectus* these two layers, the inner and outer tables of the cranial vault, are thick, and the volume of the cranial cavity varies from ca. 730 cm^3 for OH 12 (and 650 cm^3 if D2282 from Dmanisi is included in *H. erectus*) to ca. 1250 cm^3 for the Ngandong 6 (Solo V) calotte from Ngandong.

The limbs of *H. erectus* are modern human–like in their proportions (i.e., the absolute and relative lengths of the components of the limbs), but the robust long bone shafts are more flattened from front to back (femur) and from side to side (tibia) than they are in modern humans. The pelvis has a large socket for the head of the femur (the acetabulum), and the bone that connects the acetabulum to the crest of the ilium (you can feel this on yourself either side level with your hips) is thickened. Both of these features are consistent with a habitually upright posture and long-range bipedalism. Dennis Bramble and Dan Lieberman have suggested that *H. erectus* may have been adapted for endurance running. While not fast enough to catch antelopes over a short distance, these authors claim that over a long distance a persistent *H. erectus* could have eventually outpaced an exhausted antelope, much as in the fable the tortoise eventually outperformed the hare. There is no fossil evidence relevant to assessing the dexterity of *H. erectus*, but if *H. erectus* manufactured hand axes, then dexterity would be implicit.

In Africa there is evidence that later *H. erectus* may have evolved into premodern *Homo* in the form of *H. heidelbergensis*, but in Indonesia the later *H. erectus* material seems to get more specialized. This makes it less likely that the Indonesian hominins evolved into archaic *Homo* and more likely that Asian *H. erectus* was a "dead end."

Homo floresiensis

China and Indonesia (the latter especially because of the evidence from Ngandong) seem to have been among the last outposts of *H. erectus*, but all that changed with the discovery of *Homo floresiensis*. This new hominin species was established in 2004 by Peter Brown and his colleagues to accommodate a partial adult hominin skeleton (LB1) recovered in 2003 from the Liang Bua cave on the Indonesian island of Flores. The tally of fossils from the cave now exceeds 100 separately numbered specimens that are estimated to represent at least 14 individuals. The new taxon was immediately controversial for at least two reasons. First, its estimated geological age of between 12 kya and 95 kya means that it overlaps with the presence of modern humans in the region, and second, its discoverers and describers suggest that these diminutive individuals (they are estimated to be ca. 40 inches tall with a brain size of ca. 400 cm^3) sample a population that consists of a dwarfed form of *H. erectus*, an explanation that is consistent with their premodern *Homo*–like skull and limb morphology. However, a rump of researchers suggest that *no* new taxon needs to be erected because they believe the Liang Bua fossils are the skeletons of modern humans that are all affected by some growth disorder that results in small bodies with small pathological brains (the technical term is microcephalic). Both explanations are exotic, but those who espouse a pathological explanation for the individuals represented by LB1-9 have not managed to come up with a pathology that results in an early *Homo*-like cranial vault, primitive wrist and ankle morphology, and a brain that while very small, apparently has none of the morphological features associated with the majority of types of microcephaly.

Homo heidelbergensis

In Africa by 600 kya, we begin to see at sites like Bodo in Ethiopia and Kabwe in Zambia evidence of hominins which lack the characteristically horizontal and thick brow ridges seen in *H. erectus*. These crania also have a brain case whose volume averages 1200 cm^3, as opposed to the means of less than 800 cm^3 and ca. 1,000 cm^3, respectively, for *H. ergaster* and *H. erectus*. There is also a further reduction in the size of the jaws and chewing teeth. The postcranial bones lack some of the specialized features of the *H. erectus* skeleton, such as their flat shafts, but even so the limb bones of *H. heidelbergensis* are substantially thicker and stronger and the joint surfaces are larger than those of modern humans. The name *H. heidelbergensis* seems a strange one for a fossil hominin that we see first in the African fossil record, but we use it because a jaw found in 1908 near Heidelberg in Germany is likely to belong to the same taxon.

Homo neanderthalensis

The best-known species in the "premodern *Homo*" category is *Homo neanderthalensis*, better known as the Neanderthals (some researchers prefer the modern German "Neandertal," but as the name comes from the Linnaean binomial, which must retain the original spelling, "Neanderthal" is technically correct). Neanderthals are morphologically distinctive, cranially, dentally, and postcranially. Neanderthals seem to have been confined to Europe and adjacent regions, and the morphologically most distinctive later Neanderthals were subjected to sustained periods of very cold weather in what was effectively a tundra landscape.

Many researchers think that the earliest evidence of hominins that show a few signs of Neanderthal specializations come from sites in England (Swanscombe), France (Arago), and Spain (Sima de los Huesos). At the Sima de los Huesos at Atapuerca, a Spanish team led initially by Emiliano Aguirre and now by Juan Luis Arsuaga have unearthed a treasure trove of hominin fossils. These remains are approximately 400–500 kyr old and were found in a cave that was opened up when construction workers were building a new railway.

This species was given the name *Homo neanderthalensis* because the type specimen, an adult partial skeleton called Neanderthal 1, was recovered in 1856 from the Kleine Feldhofer Grotte in the Neander Valley, in Germany. With hindsight this was not the first evidence of Neanderthals to come to light, for a child's skull found in 1829, at a site in Belgium called Engis, and an adult cranium recovered in 1848 from Forbes' Quarry in Gibraltar, also display the distinctive Neanderthal morphology. No faunal or archaeological evidence from the Feldhofer cave was reported, and there seemed to be no prospect that such information could ever be obtained. However, in a remarkable example of archival research contributing to paleoanthropology, Ralf Schmitz and Jürgen Thissen managed to glean enough information about the whereabouts of the cave to go back to the much changed Neander Valley and locate the remnants of the cave sediments discarded by the miners in 1856. Excavations in 1997 resulted in the recovery of fauna, artifacts,

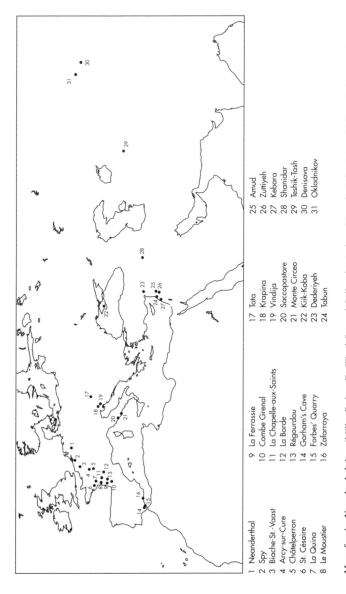

Map of major Neanderthal sites. (Miller, Barbara D.; Wood, Bernard; Balkansky, Andrew; Mercader, Julio; Panger, Melissa, *Anthropology*, 1st, ©2006. Adapted and electronically reproduced by permission of Pearson Education, Inc., Upper Saddle River, New Jersey.)

1 Neanderthal	9 La Ferrassie	17 Tata	25 Amud
2 Spy	10 Combe Grenal	18 Krapina	26 Zuttiyeh
3 Biache-St.-Vaast	11 La Chapelle-aux-Saints	19 Vindija	27 Kebara
4 Arcy-sur-Cure	12 La Borde	20 Saccopastore	28 Shanidar
5 Châtelperron	13 Régourdou	21 Monte Circeo	29 Teshik-Tash
6 St. Césaire	14 Gorham's Cave	22 Kiik-Koba	30 Denisova
7 La Quina	15 Forbes' Quarry	23 Dederiyeh	31 Okladnikov
8 Le Moustier	16 Zafarraya	24 Tabun	

and fragments of human bone and they reported "a small piece of human bone (NN 13) was found to fit exactly onto the lateral side of the left lateral femoral condyle of Neandertal 1." In 2000 more fauna, archaeological, and hominin skeletal fragments were recovered and "two cranial fragments . . . were found to fit onto the original Neandertal 1 calotte." Dates obtained from the rediscovered sediments indicate an age of ca. 40 kya for the type specimen of the Neanderthals.

After the discovery of the type specimen, the next Neanderthal discovery was from Moravia, at Sipka, 1880. Then came discoveries in Belgium (at Spy in 1886), Croatia (Krapina in 1899–1906), Germany (Ehringsdorf from 1908 to 1925), and France (Le Moustier in 1908), and Neanderthal remains have also been recovered from the Channel Islands (St. Brelade in 1911). In 1924 the first Neanderthal outside of Western Europe was found at Kiik-Koba in the Crimea. Thereafter came discoveries at Tabun cave on Mount Carmel, in the Levant, in 1929, and then in central Asia, at Teshik-Tash in 1938. In the meantime two sites in Italy (Saccopastore in 1929 and Guattari/Circeo in 1939) had yielded Neanderthal remains. Further evidence was added after the Second World War, first from Iraq (Shanidar in 1953) and then from more Levantine sites in Israel (Amud in 1961 and Kebara in 1964) and Syria (Dederiyeh, 1993). New fossil evidence for Neanderthals continues to be discovered in Europe and Western Asia, for example, at St. Césaire in France in 1979, at Zaffaraya in Spain in 1983, and at Lakonis in Greece in 1999.

Full-blown Neanderthals with all of their distinctive morphology, including a large nasal opening, a streamlined face that projects forwards in the midline, a rounded top and back of the cranium, a cranial cavity that is on average larger than that of modern humans, and distinctive

This drawing by Frantisek Kupka, with the aid of Marcellin Boule, is of the first reconstruction of a Neanderthal man, from the La Chapelle-aux-Saints Neanderthal skeleton, discovered in France in 1908. It was published in the *Illustrated London News* on February 27, 1909.

limb bones with thick shafts and large joint surfaces, are mostly found at sites that are between 30 and 100 kyr old. They sample an essentially European and Near East taxon. No Neanderthal fossils have been found in Scandinavia; it was probably too cold for human habitation. They occupied a region that during the last million years was subject to 100 kya cycles of cold weather interspersed with warmer periods.

There are two opposing views about the relationship between Neanderthals and modern humans. One suggests that they are morphologically too specialized to have made a significant contribution to the modern human gene pool, and that the differences between them and modern humans are too great for them to be included in *Homo sapiens*. The opposing view considers the morphological differences between them and modern humans to be relatively trivial and supports their inclusion in *H. sapiens*.

Mitochondrial DNA from Neanderthals

Fortunately another line of evidence is now available for assessing the taxonomy of the Neanderthals, for researchers have been able to extract short sections of mitochondrial DNA (mtDNA) from Neanderthal fossils. In their report of the first successful extraction of mtDNA from any fossil hominin, Mathias Krings and other researchers from Svante Pääbo's laboratory in Leipzig explained they had succeeded in recovering short fragments of mitochondrial DNA (mtDNA) from the humerus of the Neanderthal 1 type specimen. The sequence of nucleotides in this single fossil mtDNA sequence fell well outside the range of variation of a diverse sample of modern humans. Subsequently, mtDNA has been recovered from a second individual recently recovered from the type site (see above), from a child's skeleton from Mezmaiskaya in Russia, from two individuals from Vindija in Croatia, from the remains of a Neanderthal child from Engis, Belgium, and from one of the earliest Neanderthal skeletons to be discovered, from La Chapelle-aux-Saints in France, and from sites in Western Asia and Spain. The differences among the fossil mtDNA fragments that have been studied are similar to the differences among the same number of randomly selected African modern humans, but the differences between them and the mtDNA of modern humans are substantial and significant. The fragments of mtDNA that have been studied are short, but if the findings of these studies were to be repeated for other parts of the genome, then the case for placing Neanderthals in a separate species from modern humans would be greatly strengthened.

For a long time conventional wisdom suggested that Neanderthals evolved into modern humans. This interpretation was supported by the original dates given to a sequence of hominin fossils in the Near East. These old dates suggested that the Neanderthals found in the caves at Tabun and Amud were older than the more modern human–looking fossils from sites such as Qafzeh. However, more accurate dating methods have stood that traditional interpretation on its head. The most recent evidence suggests that the more modern-looking Qafzeh fossils predate the Neanderthal remains.

Neanderthals were one of the first, if not the first, groups of hominins to regularly bury their dead, and this is why the quality and quantity of the hominin fossil record is so much better for Neanderthals than it was for earlier hominins. Some graves show evidence of ceremony, and researchers have also claimed that Neanderthals had an interest in art.

The Neanderthals have been particularly prone to erroneous interpretations involving pathology. For example, the skeleton from La Chapelle-aux-Saints used for the extraction of mtDNA is badly affected by osteoarthritis, but it happened to be used for one of the more famous reconstructions of Neanderthals. So all Neanderthals were assumed to have a bent back and round shoulders. It was also seriously proposed that Neanderthals were modern humans affected by congenital hypothyroidism, also called cretinism. This conclusion was made on the basis of the rough correspondence between the distribution of Neanderthal sites and the contemporary "goiter belt" that extends across Europe to the Near East. But this is an example of ignoring the difference between correlation and "cause and effect." Cretinism results in distinctive marks on the skeleton that are not seen in Neanderthal fossil bones.

Neanderthal tomb burial from La Chapelle-aux-Saints, France.

. . . . .

POINTS TO WATCH

- If *H. ergaster* was the first hominin to leave Africa, it was just the first of many "pulses" of morphological and behavioral innovation that had their origin in Africa, and then spread to Eurasia, and ultimately to all parts of the world. Researchers claim that the modern human genotype retains evidence of several of these pulses, and as molecular biologists collect more information about regional variation in the nuclear genome of modern humans, more evidence may well be uncovered.

- Researchers are keen to find more sites like Dmanisi where they can gather more information about the hominin that first moved beyond Africa. Some researchers speculate that the need for a greater range associated with a reliance on meat eating was ultimately responsible for the migration. Additional fossil and archaeological evidence will allow this hypothesis to be tested by looking for evidence of organized hunting.

- Precious little is known about the origin and fate of archaic hominins like *H. heidelbergensis*. The earliest evidence for them comes from Africa, but there is very little well-dated fossil evidence from the period between 500 and 300 kya that would enable researchers to investigate how they are related to later species like the Neanderthals and *Homo sapiens*.

- Researchers are still woefully ignorant about the link between absolute and relative brain size and behavior. What were the cognitive and behavioral obstacles that needed to be overcome before hominins could rely on a steady source of high-quality foods like meat?

- Researchers are working to reconstruct the nuclear genome of a Neanderthal from Vindija. This will help clarify the relationship between Neanderthals and modern humans, and will help work out which of the differences in the nuclear genomes of modern humans and chimpanzees are specific to modern humans as opposed to hominins in general.

• • • • •

EIGHT

Modern *Homo*

•

Conventional Wisdom

For much of the last century the conventional wisdom about the origin of modern humans was that the transformation from archaic *Homo* populations to modern humans took place more or less independently in each of the main regions of the Old World, that is, in Africa, Europe, and Asia. So, for example, in Europe the Neanderthals would have evolved into European modern humans, and in Asia late surviving *Homo erectus* would have evolved into Asian modern humans. In its extreme form this multiregional hypothesis embraced the now thankfully discredited notion that geographical variants of modern humans (the term *race* has little, or no, biological meaning with respect to modern humans) were separate species with distinctly different evolutionary histories.

In this photograph taken in 1986, archaeologists continue work at the Klasies River Mouth, Republic of South Africa.

A weaker form of the multiregional hypothesis was espoused by researchers such as Franz Weidenreich (who had played a critical role in the analysis of the *H. erectus* remains from Zhoukoudian). This combined the hypothesis that regional variants of archaic *Homo* had each evolved into modern humans, with the proposal that subsequent to their independent evolution the differences between these regional variants were eventually reduced by gene flow (either by migration or by inbreeding) between the regions. Nonetheless, contemporary supporters of this weak multiregional hypothesis (WMRH) argue that despite gene flow each region has kept enough of its own character to make regional populations of modern humans distinctive and recognizable. They support the WMRH because they see morphological evidence of continuity between premodern *Homo* and modern human populations in each of the major regions of the world. For example, they claim dental and cranial evidence links *H. erectus* and modern Australians, and that a distinctive facial morphology links the Neanderthals and modern Europeans.

In this scenario for the evolution of modern humans it would be difficult to draw a line between, say, Neanderthals and early modern humans in Europe, and between *H. erectus* and early modern humans in Asia. Supporters of the WMRH argue that these gradations, together with the melding effect of the gene flow that has occurred between geographical regions, justify including *H. erectus* and all the regional hominin variants that came after it in a single species. If there were to be a single species for *H. erectus* and all subsequent hominins, then that species would have to be *Homo sapiens*, for Linnaeus's species name for modern humans has historical priority over all the other names (e.g., *H. neanderthalensis* and *H. heidelbergensis*) subsequently given to premodern *Homo* species.

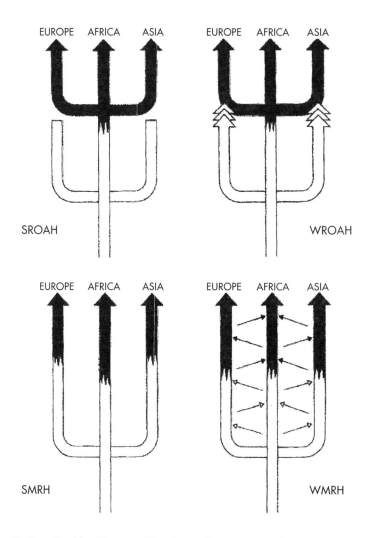

The "strong" and "weak" versions of the multiregional and recent out of Africa models for the origin of modern *Homo*. (Reproduced by permission of the American Anthropological Association from *American Anthropologist* Volume 95(1), pp. 73–96, 1993 . Not for sale or further reproduction.)

Eurocentrism in Paleoanthropology

The first discovery of a fossil modern human to be published was probably the recovery of the skeleton of what was called the "Red Lady" (it's actually a "Red Man"!) from a cave at Paviland on the Gower Peninsula, just west of Swansea, Wales, in 1822–23. However, the discovery that is nearly always cited as the first fossil evidence of modern *Homo* (i.e., *H. sapiens*) in Europe was made in 1868 at the Cro-Magnon rock shelter at Les Eyzies in the Dordogne, France. The apparent historical priority of Cro-Magnon, combined with the archaeological evidence of sophisticated small stone awls, and needles and fish hooks made from bone recovered from European sites, suggested to many researchers not only that continental Europe was the cradle of modern civilization, but that it was also the birthplace of our own genus, *Homo*, and our own species, *Homo sapiens*.

A Challenge to Eurocentrism

The preconception that Europe was the place where modern humans evolved was challenged by two developments. The first was the recognition, beginning in the latter part of the nineteenth century, and intensifying in the second quarter of the twentieth century, that there was fossil evidence of human ancestors more primitive than Neanderthals in Asia. Subsequently, of course, came the realization that the early phase of hominin evolution most likely occurred in Africa.

The second development took place in the University of Cambridge in England. It started in the 1930s with the discovery by Dorothy Garrod, a distinguished Cambridge archaeologist, of fossil remains resembling modern humans in caves on Mount Carmel in what was then Palestine. The Mount Carmel discoveries, together

Cro-Magnon shelter on the Upper Paleolithic site of Les Eyzies-de-Tayac in Dordogne, France.

with the recovery of modern human–like fossils and evidently ancient stone tools in Kenya by Louis and Mary Leakey, and in Egypt by Gertrude Caton Thompson (both also affiliated with the archaeology department at the University of Cambridge) began to convince the more outward-looking European archaeologists that important events in both the early and the later stages of human evolution may have taken place outside of Europe. In 1946 Dorothy Garrod introduced a course called "World Prehistory" into the undergraduate archaeology course at Cambridge, and her successor, Grahame Clark, continued in the same vein by encouraging his graduate students to excavate in Africa. The point of this diversion into prehistory is to make the point that by the 1950s and 1960s some students of human evolution were already comfortable with the idea that important events in the evolutionary history of modern humans may have taken place outside Europe.

Discoveries, New Dates, and Molecular Evidence

In the 1980s three lines of evidence combined to prompt some researchers to contemplate the radical proposition that Africa, far from being an evolutionary sideshow and a cultural backwater, may have been the birthplace of modern humans and of modern human behavior.

The first of the three new lines of evidence was the redating of the collections of hominin fossils in the Levant. This made it clear that instead of the Neanderthal fossils from Kebara and Amud predating the more modern human–looking fossils from Skuhl and Qafzeh, it was the other way round. The modern-looking fossils from Qafzeh were older than the fossils from Kebara and Amud that evidently belonged to an archaic *Homo* species, namely, *H. neanderthalensis*. This meant that researchers could not use dating evidence to make the case that Neanderthals evolved into modern humans.

The second line of evidence was the discovery of modern human–looking fossils in southern Africa and in Ethiopia. The most influential discovery was made in 1968 at Klasies River Mouth in South Africa. Here researchers had uncovered skull fragments that looked for all the world as if they might have belonged to a modern human, yet they were perhaps 120 kyr old. A similar date was also initially suggested for a modern human–looking cranium from a locality called Kibish in the Omo Region in southern Ethiopia. On rather weak biochronological evidence the Omo I cranium had been dated to ca. 120 kya, but a recent attempt to date the Omo I cranium using isotope dating has suggested a substantially older date, closer to 200 kya. A collection of fossils from Herto, another Ethiopian site, also suggests that modern human–like fossil hominins were present in Africa between 200 and 150 kya.

The third line of evidence came not from paleoanthropology, but from the application of molecular biological methods to the study of modern human variation. The pioneering study applying these methods was published in 1987 by Rebecca Cann, Mark Stoneking, and Allan Wilson, molecular biologists at the University of California at Berkeley. For several reasons it focused on mtDNA and not on nuclear DNA. Mutations occur in mtDNA at a faster rate than they do in nuclear DNA, and unlike nuclear DNA, mtDNA does not get reshuffled between chromosomes when germ cells divide. Nor does it have all of the innate mechanisms for DNA repair that are found in the nucleus. This may contribute to its higher mutation rate, and account for the observation that once mutations occur in mtDNA they tend to persist. The Cann et al. study compared mtDNA from 147 modern humans, 46 from Europe, North Africa, and the Near East, 20 from sub-Saharan Africa, 34 from Asia, 26 from New Guinea, and 21 Australians. The researchers found 133 different versions of mtDNA. They arranged them in the shortest tree that connected all the variants while minimizing the number of mutations. The shape of the tree they constructed from their results was striking, as was the geographical distribution of the differences between the various types of mtDNA. The tree had a deep African branch and a second branch that contained the mtDNA variants found in people from outside sub-Saharan Africa. The variation in mtDNA was not even across the tree. There was more variation within the sub-Saharan African branch of the tree than in the rest of the world put together. Not only that; most of the mtDNA variants seemed to have had an African origin.

Mitochondrial Eve

These results could mean one, or both, of two things. First, that modern humans had been in Africa longer than anywhere else in the world. Second, that the population size of modern humans in Africa was larger than that in the rest of the world combined. This makes sense, for the more people there are, the more likely it is that mutations will occur.

Cann and her colleagues made three other claims in their paper. First, because it was then widely assumed that mtDNA differences were not under the influence of natural selection (i.e., the mutations are "neutral") and because most mtDNA differences do not affect the function of the cellular machinery genes they code for, this means that any differences in mtDNA that have accumulated between two population samples are simply a function of how long those two populations have been undergoing independent evolution.

Second, Cann et al. suggested that the differences between the sub-Saharan and the non-sub-Saharan populations of modern humans would have taken about 200 kyr to accumulate, and therefore their prediction was that modern humans originated in Africa around 200 kya. Third, they claimed that the distribution of the mtDNA variants suggested that when modern humans left Africa, there was no significant interbreeding with any of the archaic populations they must have encountered as they moved into the other main regions of the Old World. Cann and her colleagues claimed that only African archaic *Homo* populations contributed to the gene pool of modern humans, and thus also they supported the corollary, which is that archaic hominins in other parts of the world made no contribution to the modern human genome. In effect, Cann and her colleagues claimed that all post–200 kyr-old hominins have only African genes.

Because you inherit the vast majority of your mtDNA from your mother, the evolutionary history of mtDNA is effectively a history of maternal inheritance. Thus, it is not surprising that either the press, or the researchers, came to call Cann et al.'s interpretation the "Mitochondrial Eve" hypothesis. It was called that because one of its implications is that the mother of all humanity was a ca. 200 kyr-old African female. I will refer to it as the strong recent out of Africa (SROAH) hypothesis, but as we will see below, most researchers who support a "recent out of Africa" model for modern human origins now support a less extreme version.

Let Battle Commence

So the battles lines were drawn: in the "red corner" the weak multiregional hypothesis (WMRH), and in the "blue corner" the weak recent out of Africa hypothesis (WROAH). Remember that some researchers who were unwilling to support the strong version of the multiregional hypothesis were more inclined to support a weaker interpretation that included gene flow between regions. Similarly, when other researchers tried to repeat Cann et al.'s calculations using more up-to-date molecular methods and more rigorous statistical techniques, they came up with different results. These still pointed to Africa as the origin of a substantial amount of modern human mtDNA variation, but several of these studies suggested there was evidence that premodern *Homo* from outside of Africa as well as from inside of Africa also contributed to the modern human mtDNA genome.

The Male and the Nuclear Perspectives

While researchers were working on ways to refine the evidence for modern human origins that could be extracted from regional variations in modern human mtDNA, other research groups had set about tackling other parts of the genome. One of the parts of the nuclear genome they paid particular attention to is the DNA from the part of the male, or Y, chromosome, which has no equivalent on the female, or X, chromosome. Because it has no female counterpart, the DNA on that part of the Y chromosome does not get reshuffled during germ cell division: the technical term for it and the mtDNA is that they are both "nonrecombining" regions of the genome. So this part of the Y chromosome DNA is like mtDNA except that it is transmitted from one generation to the next by males and not by females.

The results from studies of the Y chromosome were like those from the mtDNA studies. Twenty-one out of twenty-seven Y chromosome variants originated in Africa, and there was more variation in the Y chromosome of Africans than in all the people from other parts of the world; thus, the mtDNA results were no "flash in the pan." Much the same results have come from studies of nuclear genes, but like those in mtDNA and in the Y chromosome, nuclear gene studies are providing evidence of admixture between archaic and modern human genotypes.

The predominant message from DNA studies, be it from mtDNA, the Y chromosome, or the regular autosomal nuclear genome, is that most, but certainly not all, modern human genes originated in Africa. Another is that for the past 2 myr Africa seems to have been the source of "pulses" of hominin evolutionary novelty. The first pulse was the emigration of a *H. ergaster*-like hominin, then a *H. heidelbergensis*-like hominin, and then perhaps several waves of migration of modern human–like hominins, perhaps not looking very different, but with different cultural capacities and skills. It is now generally agreed that modern humans are derived from a relatively recent, ca. 50–45 kya migration out of East Africa. One researcher, Alan Templeton, whose important contribution pointed out the evidence for a series of migrations, gave his paper the apt title "Out of Africa Again and Again."

Migration or Gene Flow?

Novel genes can reach beyond Africa in two ways. People can take them with them when they migrate, or they can transmit them by interbreeding. The latter mechanism involves Africans interbreeding with people in an adjacent region of the Old World, these people then in turn interbreeding with other people farther away from Africa, and so on. The genes are transmitted rather like the baton in a relay race.

This is the type of gene transmission implied in one of the more recent theories about modern human origins. It is called the "diffusion wave hypothesis," and it suggests that novel genes spread in waves. It is consistent with the results of a recent study that shows a strong correlation between "genetic distance" and the actual distance in miles of the shortest overland route between where the sample of modern humans was from and the African continent.

Modern Humans Beyond Africa

There are two topics of discussion about the arrival of modern humans anywhere beyond Africa, be it in Europe, or anywhere else. One concerns the arrival of *modern human–looking people*, in other words, the earliest fossil evidence of modern humans. The other concerns the arrival of *modern human behavior*, in other words, the earliest archaeological evidence of people doing things that archaeologists are satisfied that only modern humans would have been able to do.

Table 4. The Main Morphological and Behavioral Differences Between Modern Humans and Neanderthals

	Modern Humans	Neanderthals
Morphology		
Brain size	Large	Very large
Brow ridges	Weak	Thick and arched
Nose and mid-face	Flat	Projecting
Cranial vault	Straight sides	Bulging sides
Occipital region	Round	Bulging
Incisor teeth	Small	Large
Thorax	Narrow	Broad
Pelvis	Small and narrow	Large and wide
Limb bones	Straight	Curved
Limb joints	Small	Large
Hand–thumb	Short	Long
Development— bones and teeth	Slow	Fast
Behavior		
Stone tools	Small and specialized	Larger and cruder
Composite tools	Yes	No
Shaped bone tools	Yes	No
Personal decoration	Yes, and well developed	No

Not surprisingly, the discussions about what constitutes modern human behavior are even more spirited than those surrounding what constitutes modern human morphology. Once paleoanthropologists managed to escape from the trap of equating modern human morphology with the morphology of modern Europeans, it became easier for them to recognize modern humans in different parts of the world. Archaeologists have also recognized that there is more to modern human behavior than what our ancestors were doing in Europe starting ca. 40 kya. For example, the alleged lack of cave art in Africa was sufficient to dismiss Africa as a potential source of modern human behavior. There are two good reasons to reject this argument. First, there *is* cave art in Africa; archaeologists had not been looking hard enough. Second, to have cave art you need caves, and in many parts of Africa there are no caves.

Modern Humans in Europe

The earliest fossil evidence of modern humans in Europe comes from a site in southeast Europe called Pestera cu Oase in Romania, which is dated to around 35 kya, and we know that modern human–looking people had reached England, at Kent's Cavern, by about 30 kya. The earliest evidence of modern human behavior in Europe currently comes from sites in Bulgaria called Bacho Kiro and Temnata, dated to between 43 and 40 kya, and by just less than 40 kya there are many sites across Western Europe that show evidence of modern human behavior. Modern humans in Europe overlapped with the Neanderthals for around 10 kyr or less, depending on the location. The most recent evidence for Neanderthals comes from sites such as St Césaire in France and Vindija in Croatia that are dated to ca. 30 kya.

Modern Humans in Asia: Sahul and Oceania

Researchers have suggested that modern humans may have occupied one, or more, parts of Sahul, the landmass that includes Papua New Guinea, Australia, and Tasmania, by 40 kya. With so much water locked up in polar ice caps and glaciers, land that is part of the continental shelf and which is now submerged would have provided dry connections between landmasses that are today separated by water. If hominins were in Sahul by 40 kya then they must have been in Sunda, the landmass that includes mainland Southeast Asia and the present-day islands that make up Indonesia, sometime before that.

If the late dates for the last *H. erectus* fossils in this region, from Ngandong, Java, are correct, then there would have been overlap between modern humans and late *H. erectus*. But the discovery of *Homo floresiensis*, a "dwarfed" form of *H. erectus* that persisted until 18 kya on the island of Flores, is a reminder that temporal overlap does not necessarily mean that their ranges overlapped. Different kinds of hominins could have lived on separate islands and not necessarily have come into contact with one another.

These early modern humans in Sunda must have been able to travel on rafts, or some other form of craft, and to have managed well enough to spend at least several days at sea in order to cross the open water between Sunda and Sahul. By 35–30 kya, modern humans in the Pacific region were skilled enough as seafarers to reach many remote islands in Oceania, including Timor, the Moluccas, New Britain, and New Ireland.

The existing hominin fossil record suggests that modern humans were the only hominins to enter the region we call Sahul, so there is no question of overlap with earlier groups. The time of the initial arrival of modern humans in Australia is unknown. Fossil evidence indicates that they might have arrived by 50 kya, but they were certainly there between 40 and 35 kya, when the climate was wetter than it is today.

Modern human fossils in Australia show substantial morphological variation. The people living at sites around Lake Mungo had steep foreheads, taller brain cases, and flat faces, while people at Kow Swamp and Coobool Creek in Northern Victoria had more sloping foreheads, lower brain cases, and projecting faces. Some researchers interpret these morphological differences as evidence of more than one wave of immigrants, but others see no more variation than one would expect if a new species dispersed across a large new territory such as Australasia.

Modern Humans in the New World

There were three routes from the Old World to the New, across the Bering Straits, island hopping from one Aleutian island to another, or across the Atlantic. Today all three require a sea voyage, but for several periods during the past 40–30 kyr the fall in sea level and the thick ice caused by the intensely cold conditions would have closed the Bering Straits, would have linked some of the Aleutian Islands, and would have made even a transatlantic voyage less formidable. The problem in all three cases was the intense cold those making the journey would have experienced.

The first evidence for modern human occupation within the Arctic Circle is 27 kya, and by 15 kya there is evidence of long-term occupation. During this period it is possible that modern humans following migrating herds of mammoths ventured unwittingly into the New World, but we do not find any evidence of a modern human occupation site in Alaska until 12 kya. The conventional wisdom is that the immigrants made their way south along a relatively ice-free corridor in Alaska and western Canada, and then went on to populate all of North, Central, and South America relatively rapidly. However, there is remarkably little evidence of human occupation along what is presumed to be the route south. And some New World archaeologists use this negative evidence in support of other scenarios, including one suggesting that the first occupants of the New World may have traveled there directly from Europe.

The best-known archaeological evidence for modern humans in the New World is the Clovis culture, characterized by distinctive stone tools called Clovis points. The oldest Clovis sites are dated to slightly before 11 kya, and not long after this there is abundant evidence of Clovis points over most of the unglaciated regions of North America.

Clovis points.

For a long time, archaeologists accepted the Clovis sites as the earliest evidence of modern humans in the New World. But more recently researchers have claimed they have unearthed evidence of a stone industry that is more primitive than the Clovis. The best known of these pre-Clovis sites in North America are Duktai in Alaska, Meadowcroft in Pennsylvania, Cactus Hill in Virginia, and Topper in South Carolina. In South America the best-known sites are Taima-Taima in Venezuela, Pedra Furada in Brazil, and Monte Verde in Chile. Most of these sites are dated using relative methods, but the dates of two sites, Meadowcroft and Monte Verde, are reasonably reliable. Meadowcroft's radiocarbon dates indicate it was inhabited by at least 14 kya, and perhaps as early as 20 kya.

Monte Verde provides excellently preserved evidence of modern human behavior in South America around 12.5 kya. There is even preservation of the cords used to tie hides to poles, and the remains of a dwelling that was big enough to have housed 20–30 people. Monte Verde was occupied year-round; thus it is the earliest evidence of a semipermanent occupation site in the New World.

A persistent problem with the hypothesis that the Clovis people were the first to occupy the New World is that most of the Clovis sites are in the eastern part of the United States and Canada. If the Clovis people came across what was then the Bering land bridge, how can one explain the distribution of the sites?

An archaeologist, Dennis Stanford of the Smithsonian Institution's National Museum of Natural History, has proposed a radically different hypothesis. This suggests that the first inhabitants of the New World were modern human groups from Spain. The author points out that similarities between the Iberian Solutrean tradition and some of the flakes in the Clovis toolkit support an "Iberian" rather than a "Siberian" source for the modern human settlement of North America.

It is likely there were several migrant streams of modern humans into the New World. Different groups, over different periods, arrived and settled, and each made their own contribution to the genetic and cultural diversity of New World populations. No matter when, where, and how modern humans arrived in the New World, it did not take them long to spread rapidly over a diverse range of environments. The recent announcement of the discovery of 40 kyr-old human footprints in Mexico has added yet another contentious claim to an already contentious topic.

· · · · ·

POINTS TO WATCH

- Researchers will be keen to find more sites in Africa that date to between 300 kya and the present, and to find ways of dating them reliably. Some researchers are confident that *Homo erectus* evolved into *Homo sapiens* via populations with crania like those from Kabwe in Zambia and Bodo in Ethiopia. But this may be an over-simplistic interpretation. Researchers also need to keep looking in the regions immediately adjacent to Africa for hominin evidence.

- As the technology for gene sequencing continues to improve, more genes will be sampled, and larger numbers of individuals will be sampled from each region. Researchers will be focusing on nuclear genes to see if non-African premodern *Homo* genes made a very minor, or a more significant, contribution to the modern human gene pool.

- Researchers interested in the later stages of human evolution are still unsure about the connections between morphology and behavior. Were changes in cranial shape associated with cultural changes? For example, at what stage did modern *Homo* begin to use complex spoken language, and could we tell they had reached that stage just by looking at the shape and size of the brain? Was the shift to making small, complex, stone tools the result of changes in the hands, or were these innovations entirely cognitive?

• • • • •

TIMELINE OF THOUGHT AND SCIENCE RELEVANT TO HUMAN ORIGINS AND EVOLUTION

•

5th century BCE	At least one philosopher treats modern humans as part of the natural world
1st century BCE	Lucretius suggests human ancestors were brutish cave dwellers
5th century CE	Biblical interpretation predominates
13th century CE	Thomas Aquinas reconciles Greek ideas with the biblical narrative
1543	Vesalius prepares the first detailed and accurate description of the anatomy of modern humans.
1620	Francis Bacon sets out the basic elements of the scientific method
1758	Carolus Linnaeus assembles the first comprehensive taxonomy of living organisms and establishes *Homo sapiens* as the binomial for modern humans
1800	Georges Cuvier establishes the principles of scientific paleontology
1809	Jean Baptiste Lamarck sets out the first scientific explanation for the Tree of Life
1822–23	The first fossil modern human discovery at Paviland on the Gower Peninsula, just west of Swansea, Wales
1829	Discovery in Engis, Belgium, of what later was recognized as a Neanderthal child's cranium
1830	Charles Lyell presents a scientific version of the origin of the Earth
1848	Discovery at Forbes' Quarry in Gibraltar of what was later recognized as an adult Neanderthal cranium

1856	Discovery of the Feldhofer Neanderthal skeleton
1858	Alfred Russel Wallace and Charles Darwin independently conclude that evolution is best explained by natural selection
1863	Thomas Henry Huxley presents the first science-based explanation of the close relationship of modern humans with the African apes and discusses what little fossil evidence for human evolution was known at the time
1865	Mendel publishes the results of his experiments of the inheritance of discrete traits
1864	Kleine Feldhofer skeleton made the type specimen of *Homo neanderthalensis*
1868	Fossil evidence of modern humans discovered at the Cro-Magnon rock shelter at Les Eyzies in the Dordogne, France
1890/1	Eugène Dubois discovers the first early hominin from Asia at Kedung Brubus, Java; Dubois discovers a calotte at Trinil, Java
1894	Dubois makes the Trinil calotte the type specimen of *Pithecanthropus erectus*
1907	Hominin mandible discovered at Mauer, Germany
1908	Mauer mandible made the type specimen of *Homo heidelbergensis*
1924	Taung child's cranium is the first African early hominin
1925	Raymond Dart makes the Taung cranium the type specimen of *Australopithecus africanus*
1926	Hominin teeth confirmed to be among the fossils recovered from what was then called Choukoutien
1927	Davidson Black makes one of the Choukoutien teeth the type specimen of *Sinanthropus pekinensis*
1938	Robert Broom makes TM 1517 the type specimen of *Paranthropus robustus*
1940	Franz Weidenreich transfers *Pithecanthropus erectus* and *Sinanthropus pekinensis* to *Homo erectus*
1959	OH 5 recovered by Louis and Mary Leakey; Louis Leakey makes OH 5 the type specimen of *Zinjanthropus boisei*
1964	Louis Leakey and colleagues make OH 7 the type specimen of *Homo habilis*
1968	Camille Arambourg and Yves Coppens make Omo 18.18 the type specimen of *Paraustralopithecus aethiopicus*
1975	Colin Groves and Vratislav Mazák make KNM-ER 992 the type specimen of *Homo ergaster*

1978	Don Johanson and colleagues make LH 4 the type specimen of *Australopithecus afarensis*
1979	Mary Leakey and Richard Hay describe the 3.6 myr-old fossil footprints at Laetoli
1986	Valery Alexeev makes KNM-ER 1470 the type specimen of *Pithecanthropus rudolfensis*
1989	Colin Groves transfers *Pithecanthropus rudolfensis* to *Homo* as *Homo rudolfensis*
1994	Tim White and colleagues make ARA-VP-6/1 the type specimen of *Australopithecus ramidus*
1995	Tim White and colleagues transfer *Au. ramidus* to *Ardipithecus ramidus*; Meave Leakey and colleagues make KNM-KP 29281 the type specimen of *Australopithecus anamensis*
1996	Michel Brunet and colleagues make KT 12/H1 the type specimen of *Australopithecus bahrelghazali*
1997	Jose-Maria Bermudez de Castro and colleagues make ATD 6–5 the type specimen of *Homo antecessor*
1999	Berhane Asfaw and colleagues make BOU-VP-12/130 the type specimen of *Australopithecus garhi*
2001	Brigitte Senut and colleagues make BAR 1000'00 the type specimen of *Orrorin tugenensis*
	Michel Brunet and colleagues make TM 266-01-060-1 the type specimen of *Sahelanthropus tchadensis*
2004	Johannes Haile-Selassie and colleagues make ALA-VP-2/10 the type specimen of *Ardipithecus kadabba*
2005	Peter Brown and colleagues make LB 1 the type specimen of *Homo floresiensis*
	Sally McBrearty and Nina Jablonski report the first panin fossils from Baringo, Kenya
	Draft chimpanzee genome sequence published along with the initial comparison between it and the most up-to-date human genome sequence
2009	The Middle Awash Project publishes the partial skeleton ARA-VP-6/500 belonging to *Ardipithecus ramidus* which had been recovered in a fragmentary state from a site in the Middle Awash, Ethiopia, between 1993 and 1995

FURTHER READING

•

CHAPTER 2

P. J. Bowler, *Life's Splendid Drama* (Chicago University Press, 1996): a historical account of the efforts of scientists to reconstruct the history of life on earth.

R. M. Henig, *The Monk in the Garden* (Houghton Mifflin, 2000): describes Gregor Mendel's plant-breeding experiments, and deals with how Mendel's work was rediscovered.

E. Mayr, *What Evolution Is* (Basic Books, 2001): a good introduction to the principles of, and evidence for, evolution.

J. A. Moore, *Science as a Way of Knowing* (Harvard University Press, 1993): beginning with the Greeks it traces the history of the major developments in biological research.

M. Pagel, *Encyclopedia of Evolution* (Oxford University Press, 2002): contains detailed articles about the main elements of evolutionary science.

M. Ridley, *Evolution* (Blackwell, 2003): includes both evolutionary theory and the evidence for evolution.

CHAPTER 3

J. Kalb, *Adventures in the Bone Trade: The Race to Discover Human Ancestors in Ethiopia's Afar Depression* (Springer-Verlag, 2001): focuses on the competition among scientific teams searching for early hominin fossils.

V. Morrell, *Ancestral Passions* (Simon & Schuster, 1996): describes the Leakey family and many of their important discoveries.

P. Shipman, *The Man Who Found the Missing Link: Eugene Dubois and His Lifelong Quest to Prove Darwin Right* (Simon & Schuster, 2001): describes the efforts made by Eugène Dubois to find fossil hominins in Java.

C. S. Swisher III, G. H. Curtis, and Roger Lewin, *Java Man: How Two Geologists' Dramatic Discoveries Changed Our Understanding of the Evolutionary Path to Modern Humans* (Scribner, 2000): chronicles efforts to generate absolute dates for the Javan hominins.

CHAPTERS 4–6

E. Delson, I. Tattersall, J. van Couvering, and A. Brooks, *Encyclopedia of Human Evolution and Prehistory* (Garland, 2000): detailed entries for nearly all the fossils and hominin species included in these and later chapters.

J. K. McKee, *The Riddled Chain: Chance, Coincidence, and Chaos in Human Evolution* (Rutgers University Press, 2000): argues that the evidence linking events in hominin evolution with changing climates is weak.

R. Potts, *Humanity's Descent: The Consequences of Ecological Instability* (Avon, 1997): argues that much of human evolution is a response to an increasingly unstable climate.

C. Stringer and P. Andrews, *The Complete World of Human Evolution* (Thames & Hudson, 2005): an excellent up-to-date account of the hominin fossil evidence and the methods used to interpret it.

I. Tattersall, *The Fossil Trail: How We Know What We Think We Know About Human Evolution* (Oxford University Press, 1995): a very readable account of the history of the discovery and interpretation of the hominin fossil record.

I. Tattersall and J. H. Schwartz, *Extinct Humans* (Westview Press, 2000): excellent illustrations of the hominin fossil record.

CHAPTER 7

J. L. Arsuaga, *The Neanderthal's Necklace: In Search of the First Thinkers* (Four Walls Eight Windows, 2001): the leader of the research at Atapuerca traces the rise and fall of the Neanderthals.

J. L. Arsuaga and I. Martinez, *The Chosen Species: The Long March of Human Evolution* (Blackwell, 2005): an up-to-date summary of human evolution that concentrates on the later part of the hominin fossil record.

CHAPTER 8

J. H. Relethford, *Reflections of Our Past: How Human History Is Revealed in Our Genes* (Westview, 2003): a clear and even-handed account of the implications of the interregional and interindividual DNA differences among modern humans.

USEFUL WEB SITES

http://www.mnh.si.edu/anthro/humanorigins/
This is the Web site of the Human Origins Program at the Smithsonian Institution. It is careful, up-to-date, and authoritative.

http://www.msu.edu/˜heslipst/contents/ANP440/index.htm
This is a time-space chart of hominin fossils.

http://www.becominghuman.org
This Web site is maintained by Arizona State University's Institute of Human Origins. The information is reliable and the images are carefully selected. You can see and learn about the hominin fossil record here.

http://www.talkorigins.org
This Web site summarizes the major hominin fossil finds.

http://www.sciam.com
This site has links to biographies of scientists.

http://www.ucm.es/paleo/ata/portada.htm
This site has details of the important excavations at Atapuerca in Spain.

http://www.neanderthal.de

An excellent site that features the discoveries from the Neanderthal Valley, near Dusseldorf, Germany.

http://www.chineseprehistory.org

Provides images and background to fossil hominin discoveries from China.

http://www.leakeyfoundation.org

The Leakey Foundation Web site has excellent links to other sites where readers can find information about the hominin fossil record.

http://middleawash.berkeley.edu/middle_awash.php

The Web site of the Middle Awash Project. It contains information about the recovery of the partial skeleton known as "Ardi."

INDEX

•

Note: Page numbers in *italics* include illustrations and photographs/captions.

PICTURE CREDITS

•

10.000 Meisterwerke der Malerei; 13: Carlo Crivelli, Altar of San Domenico at Ascoli, polyptych, left outer top panel: St. Thomas Aquinas/The Yorck Project: 10.000 Meisterwerke der Malerei; 50: Table of the Animal Kingdom (Regnum Animale) from Carolus Linnaeus's first edition (1735) of *Systema Naturae*/ Upload by Fastfission; 62l: Homo rudolfensis-KNM ER 1470/Upload by José-Manuel Benito Álvarez (España) —> Locutus Borg; 62r: Homo habilis-KNM ER 1813/José-Manuel Benito Álvarez (España) —> Locutus Borg; 99l: Muséum d'Anthropologie, campus universitaire d'Irchel, Université de Zurich (Suisse) : Australopithecus africanus (Sterkfontein, Afrique du sud)/Upload by Guérin Nicolas; 99r: Muséum d'Anthropologie, campus universitaire d'Irchel, Université de Zurich (Suisse) : Australopithecus africanus (Sterkfontein, Afrique du sud)/ Upload by Guérin Nicolas; 105: Olduvai Gorge/Upload by Clem23

© **BERNARD WOOD:** 3; 5; 40; 68; 80; 83; 107; 121

BRIEF INSIGHTS

•

A series of concise, engrossing, and enlightening books that explore
every subject under the sun with unique insight.

Available now or coming soon: